Galaxy Formation and Evolution

Hyron Spinrad

Galaxy Formation and Evolution

Hyron Spinrad
Professor of Astronomy
University of California
Berkeley
California
USA

Portions of the Hubble Ultra-Deep Field displayed by H. Yan and R. Windhorst. Almost every small image is a distant stellar system. Some apparent groups of galaxies are illustrated in the two "blow-ups" on the front cover.

SPRINGER–PRAXIS BOOKS IN ASTROPHYSICS AND ASTRONOMY
SUBJECT *ADVISORY EDITOR*: John Mason B.Sc., M.Sc., Ph.D.

ISBN 3-540-25498-6 Springer-Verlag Berlin Heidelberg New York

Springer is part of Springer-Science + Business Media (springeronline.com)
Bibliographic information published by Die Deutsche Bibliothek
Die Deutsche Bibliothek lists this publication in the Deutsche Nationalbibliografie; detailed bibliographic data are available from the Internet at http://dnb.ddb.de

Library of Congress Control Number: 2005924538

Printed in Germany

Cover design: Jim Wilkie
Project copy editor: Mike Shardlow

Printed on acid-free paper

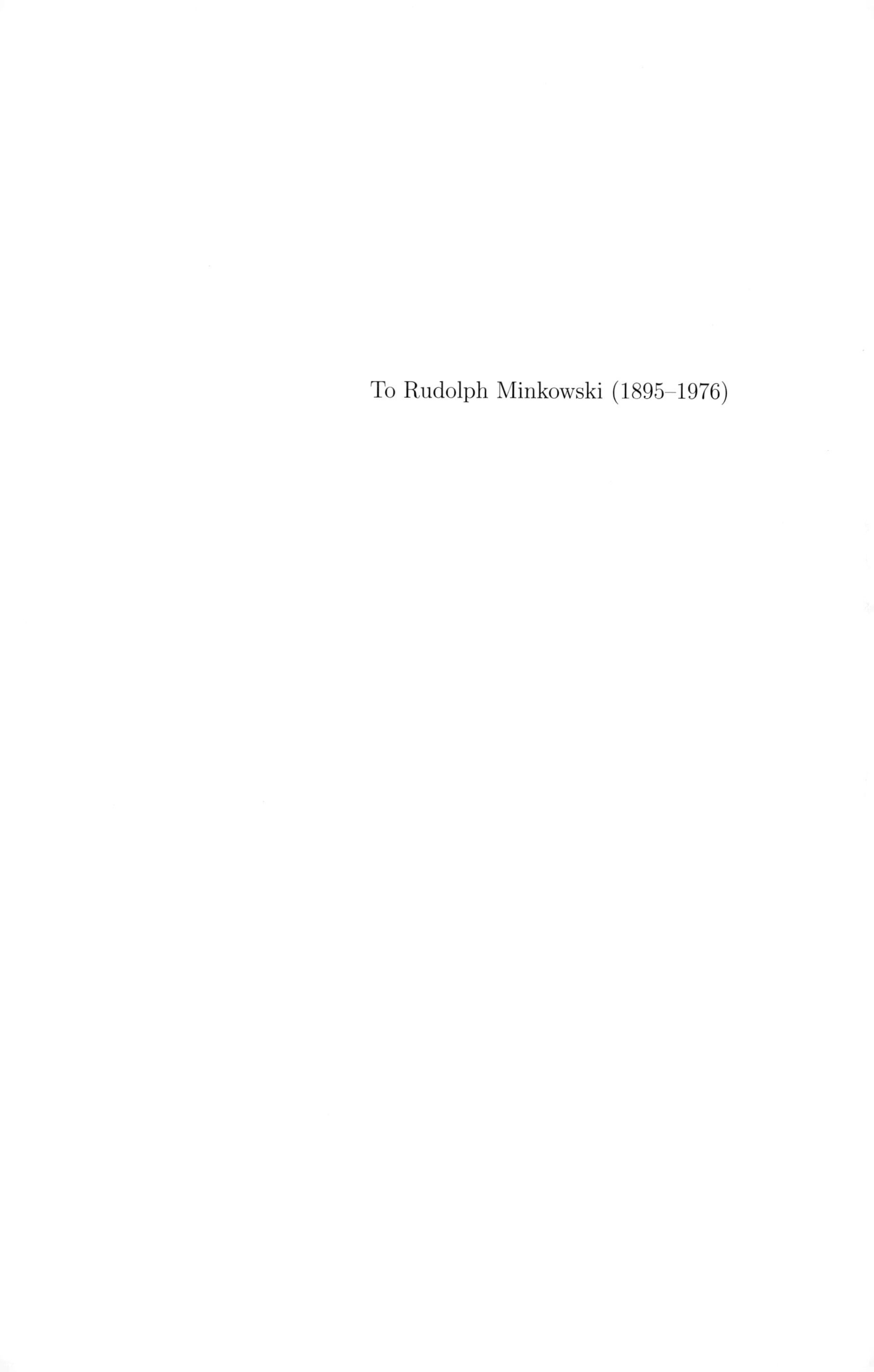

To Rudolph Minkowski (1895–1976)

Preface

An Astronomical Life – Observing the Depths of the Universe"

Though science as a subject can be difficult, what has been more important for me is that its practice can also be rewarding fun! This book is crafted to expose the reader to the excitement of modern observational cosmology through the study of galaxy evolution over space and cosmic time. Recent extragalactic research has led to many rapid advances in the field. Even a suitable skeptic of certain pronouncements about the age and structure of the Universe should be pleased with the large steps that have been taken in furthering our understanding of the Universe since the early 1990's.

My personal involvement in galaxy research goes back to the 1960's. At that point, galaxies were easily recognized and partially understood as organized collections of stars and gas. What their masses were presented a problem, which I supposed would just fade away. But fade it didn't.

Distant active nuclei and quasars were discovered in the mid-1960's. A common view of QSOs was that they have large redshifts, but what use are they for cosmology or normal galaxy astrophysics? I shared that conclusion. My expectations fell below their potential utility. In short, the Universe of our expectations rarely matches the Universe as it is discovered.

One amazing signature of the field's growth can be easily noticed by the attention given to single wave-band data, often central to extralactic studies. This approach was a historical legacy of pre-space age astrophysics which bound our view to primarily to optical wavelengths.

In 1960 galaxian astronomy was confined to the following measures,

1. optical photometry of their starlight,
2. optical spectroscopy for crude population synthesis,
3. optical redshifts,
4. hydrogen-line rotation curves (radio, 21 cm),
5. AGNs had just beeen discovered, but their connection to normal galaxies was uncertain – and a redshift controversy was emerging.

By the 1970's and especially the 1980's and 1990's, the radio detections of active galaxies became mature and accurate; thus identifications for the strong 3CR catalogue sources became practical (and eventually almost routine) despite the fact that they were associated with very faint galaxies and yielded the highest redshifts yet probed for any galaxies. False modesty aside, I developed much improved techniques to enhance the depth of redshift (distance) determinations for these distant, active galaxies, as seen in their youth (and now we'd note, their mass accumulation phases).

Since the 1980's, other wave bands have opened; one now almost routinely observes a phenomenon from γ-rays and X-rays through the UV and optical toward the near- and middle-IR. Part of the rapid development of understanding was due to instrumental improvement (CCDs were first used at large telescopes in the '80's), and part is due to the fact that powerful radio galaxies often show strong, narrow emission lines due to ionized species of oxygen, carbon and neon plus the Balmer and Lyman lines from H recombinations. These lines made their atomic and ionization state identifications, and hence their redshift, straightforwardto acquire. This effort took astronomers to $z \simeq 2$).

Of course, no single scientist can do it all. I was aided by Malcolm Longair, Simon Lilly, and, a bit later, by Wil van Breugel.

This was a heady time for me, as on good nights using the large ground-based reflectors, one could anticipate redshifts for a few distant galaxies per night – while just a few years earlier the yield would have likely been a factor of about 20 lower! Gratification was frequently rapid. My young colleagues, S. (George) Djorgovski and Patrick McCarthy enjoyed these quick results too.

A logical further step awaited us; to look at normal (as opposed to AGN and QSOs) at large redshift and thus great distance. This is difficult due to their faintness. But the rewards are greater; the use of average galaxies at a younger cosmic age is just the pedestrian comparison we need – high redshift *versus* low redshifts nearby (and more easily studied) can provide a measure of galaxian evolution in cosmic time..

This project, still ongoing, now requires the most modern CCD instrumentation on the largest telescopes and/or space imaging with the HST. A key to long-term success is the efficiency of locating distant galaxies from quality areal photometry. The HDF (Hubble Deep Field), which I am studying with younger colleagues Mark Dickinson and Daniel Stern, provides an optimally chosen corner of the observable universe in which the observing and modeling techniques can be concentrated. Collaborations that span the world concentrate these techniques and ancillary ground-based data to understand normal galaxies as they were at a fraction of the age of the current Universe. With good photometric analysis, the isolation of galaxy candidates for spectroscopic study can be almost relied-upon; we are continuing the pioneering selections first well-defined by Chuck Steidel.

Another selection technique that has led to an unexpected discovery must be mentioned here. Long spectral integrations with a narrow slit-shaped chunk of sky at high galactic latitude occasionally reveal unanticipated emission lines– some showing a few lines; some showing only one. These become unbiased detections of emission line objects, whose redshifts are straightforwardly derived. These faint galaxy spectra are sometimes dominated by a single line – the resonance line of hydrogen, Lyα at its rest-frame wavelength of 1216 Å. In such cases, spectra often show only the single emission line, with their continua lost with the background noise. Of course a large redshift is required to move a spectrum from the rest-frame ultraviolet to the observable visible wavelengths for ground-based spectroscopy. We call such discoveries serendipitously located galaxies at high redshifts. They have since been pleasently abbreviated as "SERs". I've been helped in the areal coverage by my recent collaborators, Curt Manning, Steve Dawson and Arjun Dey.

How can the chance placement of a spectroscopic slit on the sky resulting in a discovered Lyα galaxy be a useful technique? The answer appears to be astrophysically interesting; the surface density of intrinsically faint Lyα line emitting systems at high redshift must be higher than anticipated from extant surveys at moderately lower ($z \sim 3$) redshifts. These Lyα galaxies appear to be very young and incredibly small, providing an important constraint on models of structure formation. These pivotal topics for the future of extragalactic astronomy are discussed in chapters 3 and 4 of this book.

Overall, this book attempts to combine empirical discoveries and physically-based theories into a picture of the evolving universe that can be grasped by attentive fellow seekers. My hope is that it will provide to research professionals, physics and astronomy students, and active amateurs, a seasoned view of the observable Universe and it's principal components, galaxies.

Hyron Spinrad
Berkeley, California, April, 2005

Table of contents

Chapter 1

Taking the Measure of the Low-Redshift Universe

Before we begin to detail the well-understood character of galaxies nearby we wish to suggest some of the plausible and now quite conventional approaches which are astrophysically favored for the early days of galaxy evolution – their "formation".

Galaxies had to start somewhere and sometime; our conventional and partly theoretical view of their formation has to do somewhat arbitrarily with structure formation and evolution going back to the epoch of recombination (or decoupling, near $z = 1089$; Spergel et al. 2003). The structural details of $\Delta T/T$ in the microwave sky have enabled us to measure tiny density fluctuations that seed structure growth and once the Universe became matter-dominated, found their way into composite structures of dark matter and baryons. Gravity will overcome the expansion if the local density exceeds a critical value. When galaxies are seen in the early Universe, at $z \gtrsim 5$, these seeds are seen to range between small "building block" galaxies with baryons held by cold dark matter (CDM) and large proto-clusters of galaxies (Silk, 1999; Adelberger et al., 2003). Though the Universe is open to view at redshifts lower than 5, it is difficult to reconstruct how these nascent structures became what we see in the local Universe because we never see what a given object evolved from or to; we only have a "snapshot" in time. To understand galaxy evolution following recombination, astronomers have undertaken detailed comparative analyses of their properties, and coupled this knowledge with models of structure formation, and stellar and galactic evolution, using consistency with observation as the sign of success.

Our clues to the period from recombination and decoupling to the recognition of individual young galaxies ($z \lesssim 7$) and to the re-ionization of the intergalactic medium (IGM) are not yet robust. We are confident that many galaxies must grow in mass and dimensions; how they accomplish this feat is not known in detail, and indeed it might be different for differing pre-galactic clumps or atypical physical locales.

An interesting look at a plausible sequence of events is described by Adelberger et al. (2003); they take the reader from small pre-recombination acoustic waves to the epoch of early supernovae (SNe) explosions and their inherent "feedback" of energy to the IGM. Much of this book will deal with the "in-between" – from large look-back times to the present epoch.

Finally, we conclude this regression to deal with the end-products of the early structure and consequent growth and eventual physical differences that lead to differentiable classes of galactic systems. That is, the galaxy "types", here and now.

1.1 Local Galaxy Types and their Bulk Properties

Galaxies, the large and most-visible components of the main structures in our Universe, have been recognized as discrete major units of the cosmos for only some 80 years. And of course we live in one, our Milky Way galaxy (MW, often also referred to as the Galaxy). The Milky Way is probably a fairly typical, luminous spiral, so it might be advantageous to start there. Though it is intrinsically difficult to study as a basic unit of structure in the Universe (because our location in the plane of a disk-like distribution of stars, gas, and dust obscures it), the substructure is often more easily detected because of our proximity.

1.1.1 A Brief Look at the Milky Way Galaxy

The stars and gas of the MW have been studied intensively for about 100 years, and now we can categorize several types of objects/matter which dominate the mass and the physical domains of our Galaxy. Our studies in the plane of the Galaxy are still hampered by dust and gas in that thin layer. The edge-on view of the MW in the near infrared (IR) seen in Fig. 1 is a Cosmic Origins Background Explorer (COBE) view of our Galaxy by Mike Hauser and the COBE collaboration. Our vantage point is at a galactocentric distance of about 8 kpc from the center. At the near-IR wavelengths, the light is dominated by evolved red giants and subgiant stars; this illustration shows well the relative magnitudes of the two main stellar components of the MW: the disk (Population I), and the bulge, which is quite bright, but diminutive in size. Most of the stars seen above and below the plane of the disk are more local stars within the disk. The disk can be separated into three components, a very young population which are located in the spiral arms, a thin disk which has been smeared out by differential rotation, and a thick disk. The thick disk is thought to be older, and thick due to stellar interactions with giant molecular clouds. The third main component is the low-density stellar halo known as Population II stars. There is also a fourth component, optically invisible, but dominant in mass, which is the dark halo. Values for the Milky Way parameters are given in Table 1. The disk appears to be old, but the younger disk stars have a smaller scale-height than the old stars. The bulge stars may be predominantly contemporaneous with the Population II stars. However, they are much more concentrated, and their metallicity is much higher. In addition, there is a strong metallicity gradient toward higher values near the center. The bulge has a mean prograde rotation rate of about 100 $\mathrm{km\,s^{-1}}$, while the halo has essentially no net rotation. Thus, roughly half of the halo stars have retrograde orbits. Both the stars and the globular clusters of halo stars have plunging orbits, while Population I stars are generally circular.

Interestingly, the galaxy disk stellar mass ($M_{\mathrm{disk}} \simeq 6 \times 10^{10}$ $\mathrm{M_{\odot}}$) is much higher than the current gas mass M_{gas} by a factor of about 10, and is less than the dark halo mass inferred in the MW from its flat rotation curve, and its radial extent to at least 20 kpc where a total dark mass of 3×10^{11} $\mathrm{M_{\odot}}$ is inferred. This non-luminous mass is generally assumed to be slightly ellipsoidal in shape, though the degree of flattening is not well-constrained.

FIGURE 1: The Milky Way, seen in the 2-3 μm near-IR with the COBE satellite. Image courtesy of Mike Hauser and the COBE collaboration. See text for discussion. (Reproduced in colour in the colour section.)

Our galaxy nucleus is perhaps a bit atypical; besides the likely fairly large black hole (BH) – a mass of some $\sim 3.7 \times 10^{6}$ $\mathrm{M_{\odot}}$(Genzel et al., 2003; Ghez et al., 2003). The observed peculiarity is a cluster of B-stars near the center (within one parsec); most spiral galaxies with bulges do not have an associated young stellar population near their nuclei (see Spinrad & Peimbert 1975). The overall bulge population in the MW and other galaxies is predominantly an old population with fairly luminous cool stars dominating the light.

In the 1960s and 1970s, astronomers continued their efforts to correlate stellar properties with structural units in the MW; in particular, the characteristics of stellar populations – star ages and abundances of heavy elements, and the MW kinematics and distributions – correlate to an impressive degree with the identified stellar populations (Pop. I and Pop. II and their sub-components). Overall, the more recent efforts to expand on the correlations shown in Table 1 have become more complex and varied. A close-up view of any normal spiral galaxy would likely be the same; our attention to galaxies beyond the MW (to "local" types of galaxies) will follow in this chapter.

TABLE 1: Characteristics of MW Populations

Population name	Extreme Pop. I	Older disk pop.	Thick disk	Bulge –	Halo (stellar) (Pop. II)	Halo (dark)
Typical	ISM:H,H_2,CO	Sun	old disk stars	very old stars	Metal-poor	dark matter
constituents	O-B stars	most nearby stars	old stars	some young stars	high-velocity stars	–
–	Stellar births	older open clusters	–	–	"retrograde" stars	–
–	Cepheids	–	–	–	RR Lyrae ($P > 0.5^d$)	–
–	–	–	–	–	globular clusters	–
z-width	$\sim$ 100 pc	200 pc	400 pc	$\sim$ 1 kpc	$>$ 20 kpc:	$\gtrsim$ 200 kpc
Age ($\times 10^9$)	< 0.1	0.5 to 5	$\gtrsim 6$	$\gtrsim 10$	$\gtrsim 10$	–
Conc. to Gal center	slight	moderate	moderate	strong	moderate	(?)
Abundances [Fe/H][a]	0.0 ± 0.2	$0.0^{+0.1}_{-0.2}$	-0.5	0.0 to 0.6	-1.0 to -3.9	–
In spiral arms	yes, usually	partly	no	no	no	–

Notes: a $[Fe/H] \equiv \log(Fe/H) - \log(Fe/H)_\odot$.

1.1.2 The Hubble Classification Scheme

Scientists faced with a new subject or a new phenomenon often find it of value to establish some sort of empirical classification scheme to separate the "new" objects with simple or readily available parameters that may or may not be impersonally measured. The Hubble classification scheme for galaxies was begun in the 1920s and 1930s, and initially published by Edwin Hubble in 1936. While this work is, on its surface, a cataloging of the apparent shapes of galaxies, their morphology turns out to have a strong, if poorly understood relationship with the physical content of stars and gas.

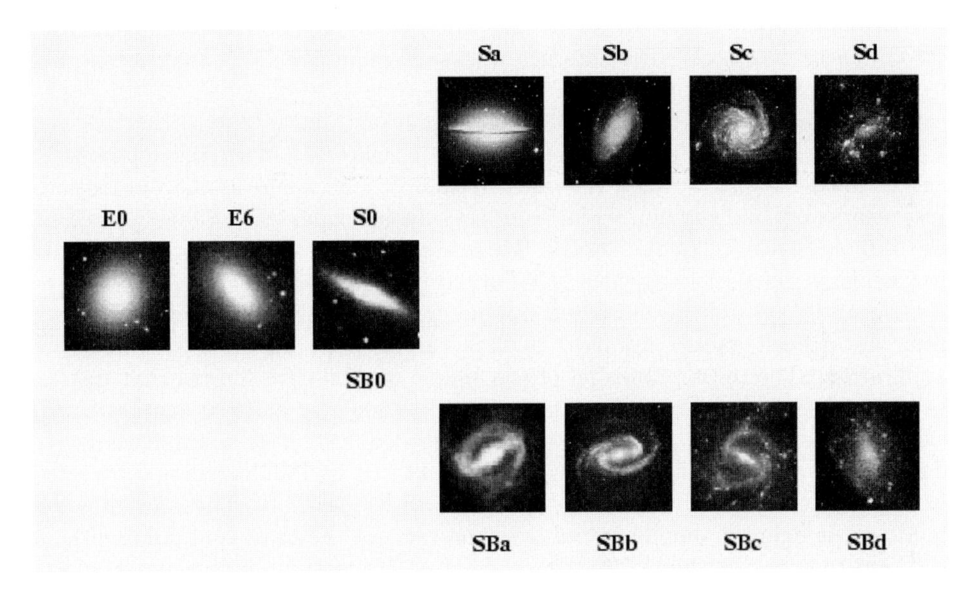

FIGURE 2: The famous Hubble tuning fork diagram; the illustration is by Ray White http : //www.astr.ua.edu/preprints/white/gal_tuningfrk.html. "Early" galaxies are on the left, and "late" galaxies are on the right. Note the decrease in galaxy central bulge/disk ratio as one progresses to the right in this figure.

The Hubble classification of spiral galaxies, in particular, places them in a robust order of three parameters. The first, in modern astrophysical language, would now be called the galaxy bulge/disk ratio; the second is the degree of resolution of stars and ionized hydrogen (H ii) regions in the arms, and the third is the "openness" of the pitch angle of the spiral structure. These three parameters are strongly correlated, and also differentiate the Hubble classes [Sa, Sb, Sc] on the basis of color, mean equivalent stellar spectral type, gas content, and to a lesser degree, the total luminosity. Indeed, astronomers have attempted other classification procedures, qualitative to semi-qualitative, that are also correlative to more physical measures. The subjective schemes invented by de Vaucouleurs (1959) and by van den Bergh (1960b,a), and by Morgan (1958) are less widely used by contemporary astronomers, but despite their "speciality in classification", these schemes still correlate with certain integrated physical properties of the galaxies, such as overall color or even their star-formation rates.

In particular, the de Vaucouleurs extension of the Hubble spiral galaxy portion of the famous "tuning fork" diagram to types Sd and Sm was an obvious addition to Hubble's system of galaxy classification; it also represents a trend of bluer integrated colors that correlate with the "lateness" on the spiral sequence (toward the fork (right) in the tuning-fork diagram).* But as a sign for classification "success", a majority of nearby systems can be easily placed within the "large boxes" of the well-tried Hubble classes. A very literal version of the tuning fork is seen in Fig. 2. The reader is also referred to the beautiful prints of nearby galaxies in Sandage & Bedke (1975) and *The Hubble Atlas of Galaxies* (Sandage, 1961). Those with convenient internet access can enjoy the CCD images collected by Z. Frei and J. E. Gunn on the Princeton website at
http : //www.astro.princeton.edu/ ~ frei/Gcat_htm/Catalog/Jpeg/

In the context of this book we are mainly concerned with the relations of galaxy morphologies between the "nearby sample", as found in the illustrations just mentioned, and the morphologies of very distant galaxies, perhaps imaged by the Hubble Space Telescope (HST). These distant systems are obviously harder to classify as our best images are still limited in spatial acuity.

It may be possible to employ some impersonal classification indices to both near and distant galaxy samples; to date this concept has had only marginal application. The most promising of the simple quantitative indices that distant galaxy morphology might encapsulate are the asymmetry and the concentration. Our concepts of these characteristics can be encoded into image manipulations. To determine the asymmetry index, A, an image rotated by 180 degrees is subtracted from the true image, and residuals compared to the original. The concentration index, C, is defined roughly in terms of the ratio of the radii containing 80% of the light to that containing 20%. We expand on this system in Chapter 3.

The values of both indices correlate with the Hubble type, basically, according to the bulge to disk ratio. The concentrated E-galaxies can be numerically characterized by the C index to moderate cosmological distances, and thus this method is relevant to our review (see Bershady et al., 2000; Abraham et al., 1996; Conselice et al., 2003).

1.1.3 Morphological Breakdown of Nearby Galaxies

Returning to our "broad-brush" look at the Hubble classes of galaxies, the differences in the bulge/disk photometric ratios is striking as we move from E galaxies to S0s to big bulge spirals (Sa, Sb) to galaxies with almost unrecognizable central brightness peaks (and thus small bulge/disk ratios). A virtue of the Hubble system is its large bin steps, and that makes classification through the normal spiral sequency easy. The S0 galaxies on the other hand are quite difficult to differentiate from the Es, and even some Sa systems.

---

*Generally, elliptical and lenticular (S0) galaxies are colloquially deemed "early", and spirals and Irr galaxies are "late".

The nearby Universe yields quite interesting information on the galaxy types and their local environments. In the low-density field (Longair, 1998; de Vaucouleurs, 1963) bright galaxies have a morphological breakdown as shown in Table 2.

TABLE 2: Morphological Breakdown of Luminous Galaxies in the Field (a low-density environment) (Dressler, 1980)

Type	Fraction
Spirals	61%
E	13%
S0	21%
Irr	3%

In regions of high density a dramatic change is found: S0 and E galaxies now dominate, as shown in Table 3, with data from Dressler (1980) found. This variation in morphological breakdown seen between the field and clusters is now known as the morphology–density relation between the ambient galaxy density and their morphology. One might suppose that in denser environments, many of the evolutionary forces, such as merging, or tidal effects, are speeded up in dense environments.

TABLE 3: Morphological Breakdown of Luminous Galaxies in Dense Environments (Dressler, 1980)

Type	Fraction
Spirals	11%
E	41%
S0	48%
Irr	–

As a short aside, we mention the obvious – since galaxies are spatially extended and have no well-defined "edge", there is always a limitation of the photometry of nearby galaxies – and the integration of their surface brightness distribution. At the low surface brightness encountered at large galactocentric distances, the galaxy light competes poorly against the brightness of the foreground sky (in the visible/red for ground-based observing, and even in the 5 μm region for space-based observations). Luckily, new space efforts, like the Spitzer IR satellite, have improved the situation at intermediate (around 5 μm) IR wavelengths.

1.1.4 Surface Brightness Distributions

The surface brightness distributions for nearby galaxies are roughly bi-modal. Accordingly there are two types of adequate functional fitting-formulae astronomers have used to parameterize and extrapolate the galaxy surface brightness distributions.

For E galaxies and the larger bulges of Sa/Sb spirals, a power-law, such as in Eq. (1) is useful:

$$I_\lambda = \frac{I_{0,\lambda}}{\{(r/r_0) + 1\}^2}, \tag{1}$$

where I_0 is the central surface brightness, and r_0 is a characteristic interior radius (where the surface brightness falls to 1/4 its central brightness). This is the well-known "Hubble brightness law". Another representation is the de Vaucouleurs "$r^{1/4}$–law" for E-systems,

$$I(R) = I_e \exp\left\{-7.67\left(\left(\frac{R}{R_e}\right)^{1/4} - 1\right)\right\}, \tag{2}$$

where I_e is the surface brightness at R_e, and R_e is the effective radius; the radius of the isophote containing half of the total galaxy luminosity. For the disks of spirals (Freeman, 1970) the formulae found useful by trial and error is an exponential function,

$$I_D = I_0 \exp\left(-r/h_r\right), \tag{3}$$

where h_r is a scale-length, and I_0 is the surface brightness extrapolated to the center. I_0 is the intercept surface brightness of the disk extrapolated to the galaxy center. This yields a (mostly) linear plot if surface brightness is in magnitudes. These fitting-formulae are, of course, simplified and general, but give us the concept of a galaxy's run of star density with galactocentric distance.

If our primary goal is to study the distribution of local galaxy light and mass, and its integral (the total galaxy luminosity, or M_V, say), we will use the fitting-formulae, but even more critical will be the distances to individual systems and the extragalactic distance scale in general.

Indeed, the problem of the extragalactic distance scale has loomed threateningly against the pursual of several astrophysical goals in the preceding decades. In any absolute luminosity measurement, the distance scale (often simply defined by the extant Hubble constant, H_0) is usually more critical than the photometric methods to cover the galaxies' luminosity (integral).

1.1.5 The Colors of Local Galaxies

In conclusion of this introduction to the photometry of local galaxies, we list in Table 4 the averaged broad-band colors of galaxies of different morphologies. The capital letters indicate, blue (B), 2 μm (K'), visual (V), and earth-based ultraviolet (U) bands/wavelengths. Note the large range of "blueing systems" as we go to "later" classes, especially in the $(U - B)$ color. Further information on nearby normal galaxy colors may be found in references such as Coleman et al. (1980) and de Vaucouleurs (1961).

TABLE 4: Galaxy Colors as a Function of Morphological Type

Type	$(B-K')$ (mag)	$(B-V)$ (mag)	$(U-B)$ (mag)
E	–	0.96	0.49
S0	4.0	0.91	0.42
Sa	–	0.87	0.31
Sb	3.6	0.77	0.18
Sbc	–	0.71	0.08
Sc	3.1	0.62	0.00
Sd	2.6	0.58	-0.09
Sm	2.5	0.42	-0.28

1.1.6 Distance Calibration

Galaxy distances are important desiderata; over the last year or two, several lines of direct and indirect evidence have greatly lowered the uncertainty in the Hubble constant, H_0. The Hubble constant is often represented by the variable h, defined as

$$h = \frac{H_0}{100}. \tag{4}$$

For this book we use the currently preferred value $H_0 = 70\ \mathrm{km\,s^{-1}\,Mpc^{-1}}$. Thus we employ the value, $h = 0.7$. This enables us to establish a self-consistent distance-scale for many moderately distant (but *not* local or at cosmological distances) galaxies,

$$d = \frac{v}{H_0}. \tag{5}$$

For relatively nearby systems there is now fair convergence on the distance determination criteria. These are currently based upon similar luminous star comparisons, "normalized" within the Milky Way galaxy. The standard candle of choice is often a Cepheid Variable. The key galaxy to utilize as an intermediary is the Large Magellanic Cloud (LMC); however, we do note that the Cepheid distance to the LMC doesn't agree perfectly with a scale derived from comparisons of the shorter-period variables, the RR Lyrae stars, ubiquitous to all galaxies where they can be seen. The modern distance moduli to the LMC with these varied criteria are mentioned in the Binney & Merrifield (1998) text; we repeat their individual distances to the LMC in Table 5.

A pessimist might view the discrepancy between the distances by the assumed best yardsticks as a substantial disclaimer on the extragalactic distance scale. It is certainly a mild worry for nearby extragalactic numerology.

TABLE 5: LMC distances by Different Comparisons

Star and Technique	D (kpc)	References
(Upper) Main Sequence fit	50 ± 5	Walker (1992)
Cepheid P-L	50 ± 2	Feast & Walker (1987)
RR Lyrae ($M_v = 0.7$)	44 ± 2	Walker (1992)
SN 1987a; (Baade-Wesselink)	55 ± 5	Branch (1987)

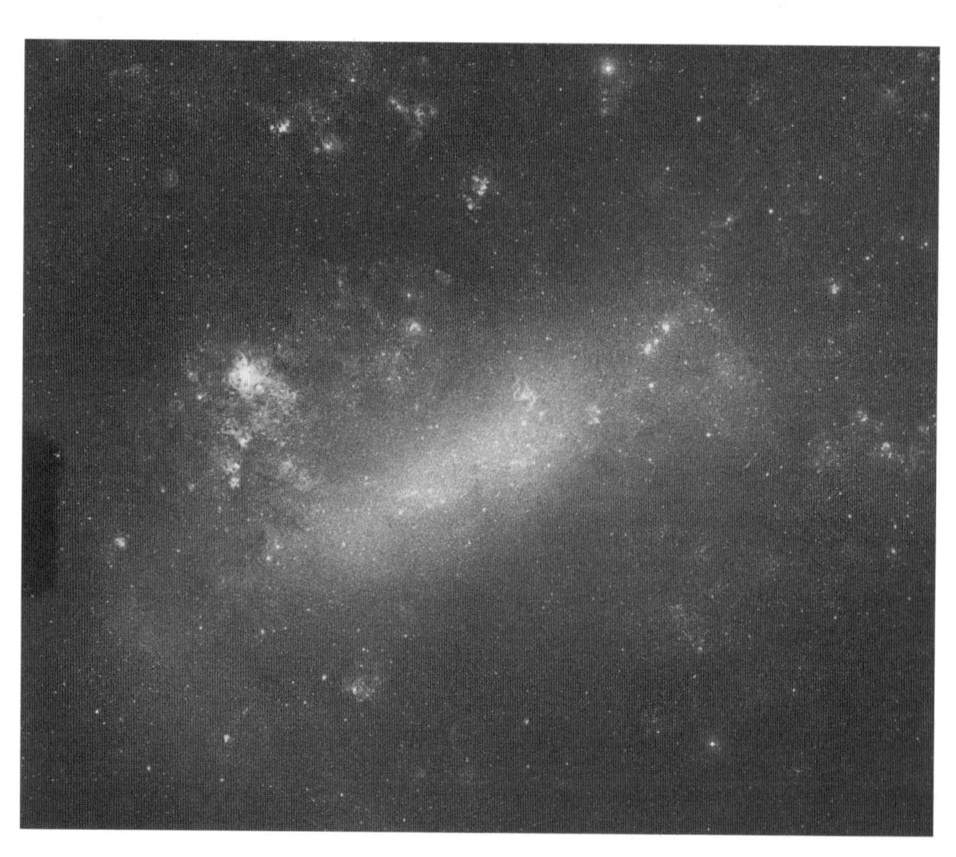

FIGURE 3: The Large Magellanic Cloud, an irregular galaxy beginning to merge with the Milky Way galaxy. This "almost Irr" galaxy is close enough as to be a valuable yard stick in the search for a consistent extragalactic distance scale. Image taken with the Anglo-Australian Observatory's Schmidt camera; courtesy of Anglo-Australian Observatory/David Malin Images. (Reproduced in colour in the colour section.)

1.1.7 The Luminosity Functions of Local Galaxies

It is often of considerable interest to describe the range in the brightness of present-day galaxies over an adequately defined volume. The space densities of galaxies of different luminosities (absolute magnitudes) is called the galaxy luminosity function, usually plotted as the log number of galaxies per magnitude interval per cubic megaparsec. Unfortunately we do not have a very robust *mass* function to accompany the luminosity-based relation.

Locally, we can establish the luminosity function either from counts of rich cluster members, so that the bins in absolute magnitude are well-filled, or by careful comparison of catalogued galaxy luminosities, where the pitfalls of technique tend to be statistical completeness worries (and possible bias).

In recent years the galaxy luminosity function (GLF) has been assumed to fit well a functional representation suggested by P. Schechter (Schechter, 1976). It has the form,

$$\Phi(L)\,d(L) = \Phi_0\,(L/L^*)^\alpha\,\mathrm{e}^{-L/L^*}\,d(L/L^*). \quad (6)$$

Here, $\Phi(L)$ is the number of galaxies in the range $L \pm \Delta L/2$ per cubic Mpc, L is the galaxy luminosity (linear), Φ_0 is the normalization factor of the GLF (requiring a volume-limited sample), α is the slope of the power-law observed on the faint end of this function. $L = L^*$ is the luminosity position in the functional shape corresponding to the "bend" of the analytic function; observationally, L_* is not terribly well-observed and thus not sharply defined (it is about 2×10^{10} $\mathrm{L}_\odot$). In magnitude units, the corresponding $M_*(R)$ is -21.0 or so with $h = 0.70$. Most of the galaxy LF light comes from galaxies near L_*; the brighter ones are rare. Note that when $L = L^*$, $\Phi(L) = \Phi^*/e$.

The Schechter LF is also often presented in terms of galaxy absolute magnitude,

$$\Phi(M)\,dM = 0.4\log_e 10\Phi_*\left[10^{0.4(M_*-M)}\right]^{1+\alpha}\mathrm{e}\left[10^{0.4*M_*-M}\right]dM, \quad (7)$$

where M is the galaxy absolute magnitude. Figure 4 shows the Schechter LF fit for cluster galaxies, where the faint-end slope is $\alpha = -5/4$.

The Schechter function is a rather good overall fit to the GLF observations (De Propris et al., 2003). However, if one dissects the luminosity function by morphological types (Binggeli, 1987; Marzke et al., 1998), the results are not particularly pleasing, as the Irregular and late Sd/Sm galaxies include no luminous systems and a quite steep faint-end slope, while the Sa and Sb systems, as well as the local E/S0 galaxies do include galaxies up to $M_B = -22^m$ ($M_V = -23^m$), and fewer intrinsically faint ones, leading to a flatter faint-end slope.

A useful local by-product of the calibrated GLF is the spatial integral over Φ; in total luminosity units it reaches $L = 1.4 \times 10^8\,h$ $\mathrm{L}_\odot\,\mathrm{Mpc}^{-3}$. This is, then, the luminosity density of our nearby Universe. How does it vary with cosmic epoch? We will deal with this interesting problem in the ensuing chapters of this book.

1.1.8 Gas in Nearby Galaxies

Among several potentially relevant comparison themes we may choose to detail the cosmic evolution of the galaxy gas content. The interstellar medium (ISM) content in galaxies has wide variations. We now can measure molecular gas, neutral H gas, and ionized species. We tackle these phases in reverse order.

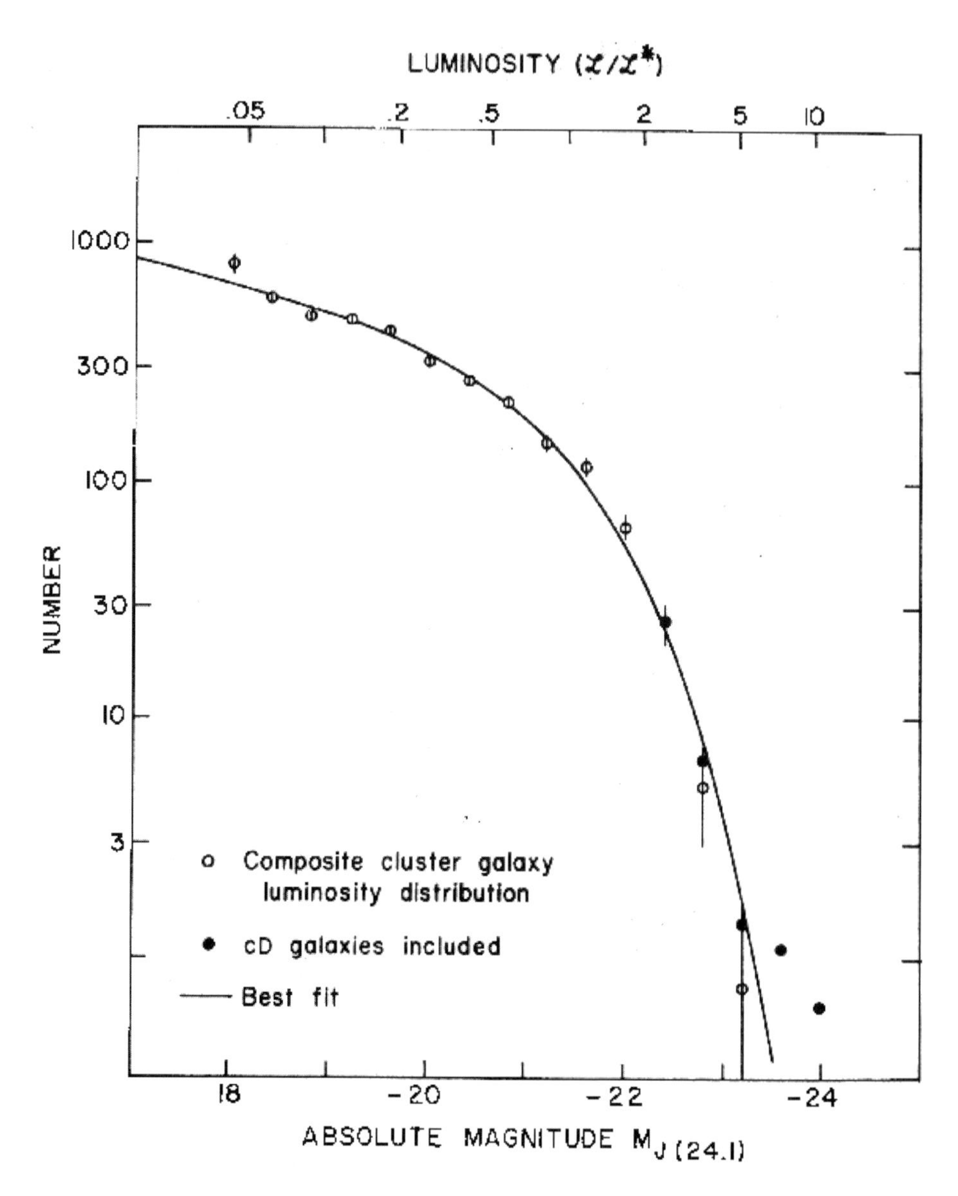

FIGURE 4: The Schechter GLF for cluster galaxies from Schechter (1976). Absolute magnitude bins as well as luminosity are shown. For the adopted Hubble constant $H_0 = 50$ km s^{-1} Mpc^{-1}, the absolute magnitude of an L^* galaxy is $\simeq -21.5$ mag (or -20.6 mag for h = 0.7) in B. The faint-end slope parameter is $\alpha = -1.25$.

The presence of at least a modest amount of ionized atomic gas in all morphological types of galaxies has been known for decades. The detailed local distributions of the well-known optical lines [O II]λ3727, [O III] 5007, 4959, Hα (6563), and

Hβ (4861) were qualitatively understood by the 1960s; to zeroth order the galaxies with the hottest blue giants and main sequence OB stars had the strongest and most consistent optical emission lines. The E/S0 galaxies with (mainly) red stars rarely showed emission, occasionally high-quality observations of E systems (high S/N and moderately spectral purity) show a weak [O II] signature, often confined to the inner regions of the galaxy.

The modern literature on the ionized gas consistently argues for a modest or negligible amount of this gas phase in the ISM of all galaxy types. But it has considerable import for light element abundance studies, and occasionally, for the study of weak sources of ionization,. And especially we note the presence of ionized gas around the hot stars within the ragged boundaries of H ii regions in late-type (Sc, Sd, Sm) galaxies. In the context of this book the strengths (the equivalent widths, actually – see Kennicutt, 1992) of the stronger emission lines here and now in the local star forming Universe are modest. They clearly evolve with cosmic epoch, so we shall face this situation explicitly in a later chapter.

Much of the mass and extent of the ISM in normal galaxies is due to the presence of neutral gas, mainly H II. Following Roberts & Haynes (1994) we note that the total H I mass (in solar units) is calculated to be

$$M_{\rm H\,I} = 2.36 \times 10^5 D^2 \int S\, dV, \quad (8)$$

where D is the galaxy distance in Mpc, and $\int S\, dV$ is the H I 21-cm line flux in Jy kms units. It is also convenient at times to calculate for comparisons the H I surface density ($M_{\rm H\,I}/\pi R^2$) where R is the optical linear radius. Another useful ratio is $M_{\rm H\,I}/L_B$ (the H I mass per unit blue band luminosity).

When we examine the global neutral H gas content we find a rich statistical sample as $> 10^4$ galaxies have H I fluxes; the morphological type correlation with H I total mass is uniform with type, and the H I$/L_B$ ratio is strongly correlated with morphology, also. Table 6, below, mainly from a "smoothed" Roberts and Haynes listing, indicates the well-known trends between gas and morphologic class.

TABLE 6: H I versus Morphological Type

Parameter	E/S0	S0/Sa	Sb	Sc	Sd	Sm
$M_{\rm H\,I}$ (10^9 M$_\odot$)	1.0	4.	10.	11	7.	2.
$M_{\rm H\,I}/L_B$ (M$_\odot$/L$_\odot$)	0.04	0.11	0.21	0.29	0.40	0.70
$M_{\rm H\,I}/M_T$	0.003	0.03	0.05	0.08	0.11	0.15

The main point here is that E/S0 systems locally are rather devoid of neutral H gas, but not completely so. The small amount of H I could be from a primeval infall, it could be partly from stellar mass-loss, and/or it could be the result of a merger with a (small) gas-rich system.

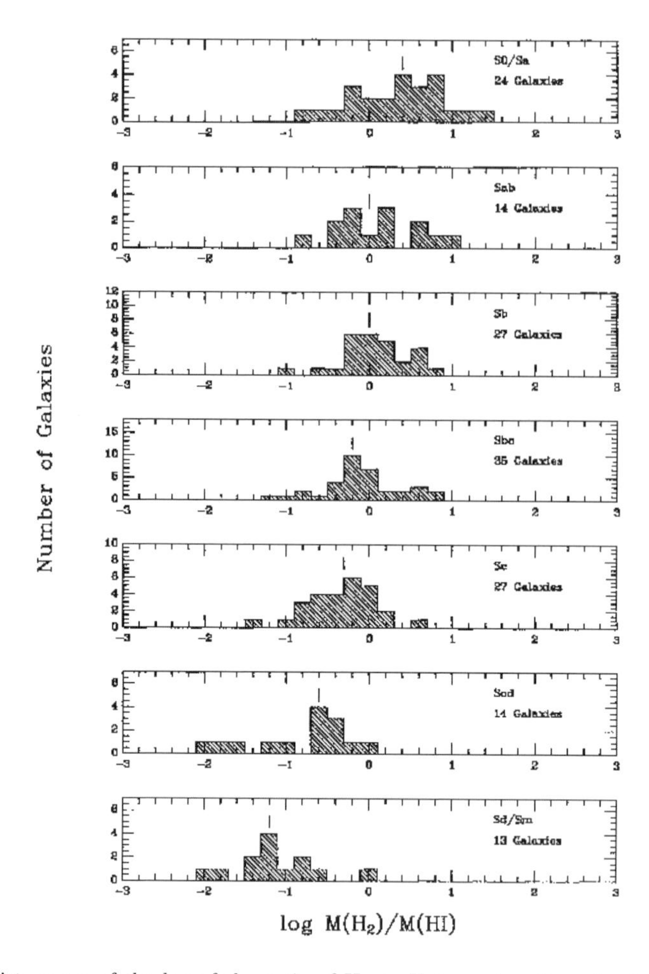

FIGURE 5: Histograms of the log of the ratio of H_2 to H I as a function of galaxy type. Note the gradual shift to a smaller proportion of molecular gas in the "late" systems (Sc, Sd, Sdm types) toward the bottom of this figure. Illustration from Young & Scoville (1991). Reprinted, with permission, from the Annual Review of Astronomy and Astrophysics, Volume 29 (c)1991 by Annual Reviews www.annualreviews.org

The situation for molecular gas is somewhat more "evolving" and indirect, as the main observable molecular lines are due to CO, the nominal alias for all galactic molecules. H_2 – the presumed main molecular component – is hard to observe, as it is a homonuclear molecule, and requires observation of IR quadrupole lines with low-transition probabilities. Figure 5 shows the ratio of the masses of H_2 to H I for the range of spiral galaxies. Though the range of H_2/H I is significant for each galaxy type, the trend in the mean value clearly decreases for late-type spirals.

1.2 Starbursts and ULIRGs

Galaxies of types Sa–Sc have the largest fraction of molecular hydrogen. The latest systems seem to show less CO; there are considerable uncertainties in CO/H_2 observations in E/S0 galaxies. But weak CO has been detected in a few bona fide E galaxies; cool gas is present, although amounts are qualitatively small.

It is, of course, well documented that the ISM (neutral + molecular) is largely "pre-stellar stuff". So when we look back in space and time at more youthful galactic systems, their ratio of gas/stars may well be systematically different than observed locally. However, it no longer seems, as it once did, that galaxies are "co-evolving": co-evolving galaxies of a given type would have been formed close to a given time, and form stars at a given time, and become mature at another given time. Possibly because of the wide range of environments, it now seems as though galaxies of any given type may have been formed over a wide range of redshifts, or may be developing from one type into another over cosmological time scales. However, as we analyze the low-redshift universe, we find the environmental dependencies specific to types of galaxies may often be characterized in intuitively straightforward ways. For instance, many starbursts are associated with gas-rich locales.

Starbursts, scenes of rapid star-formation (SF), often sampled best at IR wavelengths, occur in fairly small, high-surface brightness regions in galaxies, dominated by (young) starlight, but often are partly (or largely) obscured by dust. Starbursts at a distance attract us by their associated IR emission. The dust presumably re-radiates the absorbed UV as a strong IR flux – this re-processing is the best physical signpost for recognition of a starburst, often important at $\lambda \gtrsim 25\,\mu$m.

A classical example of a starburst region in a nearby well-resolved galaxy is M82. M82 is superficially an irregular galaxy of modest visual luminosity some ~ 5 Mpc from us. The Infrared Space Observatory (ISO) has confirmed substantial emission from dust in the mid-IR ($11 \lesssim \lambda \lesssim 15\,\mu$m), attributable to very small dust particles (Schreiber et al., 2003), probably dust at temperatures $T = 150$ to 200 K (for an emissivity index $\epsilon = 2$). The star-formation rate can be estimated from the UV luminosity, but since it is often highly extincted, estimates may be inaccurate. In this case, the IR flux can be used as an indicator of the star-formation rate. The emission appears to come from within the optical contours of the starburst of size $\lesssim 1$ kpc. The observation of compact radio sources and subsequent imaging of expanding shell-like structures suggests that the star-formation rate can also be determined by the number and size distribution of SN remnants (Pedlar et al., 2003).

Starbursts can also occur during mergers of spirals. According to the theoretical models, during the initial stage, starbursts can be formed during the initial stages of coalescence (Kewley & Dopita, 2003), and form forming super star clusters, such as in the "antennae" galaxies, NGC 4038/9. During the final stages, when the two nuclei are coalescing, nuclear starbursts may result. Chandra, a new X-ray space telescope, has helped to find these merging active galactic nuclei (AGN).

Extreme cases of starbursts illustrate the expelling of huge quantities of dust from the rapidly evolving massive stars. These giant dust clouds absorb the energetic photons (extinction is much more effective in the UV than the IR), and re-radiate mid- to near-IR photons, producing what is known as a luminous infrared galaxy (LIRG), or an ultra-LIRG (ULIRG). Nevertheless, energy must be conserved, so what is hidden by dust may be gleaned from the analysis of the combined output of the galaxy in all spectral windows.

The relatively small fraction of galaxies undergoing this extraordinary level of star-formation and energy production suggests that these are relatively short periods in the life of the galaxy. What is the photometric fate of these remarkable galaxies? The picture is not perfectly clear, but what we have learned suggests the following rough scenario. Starbursts, whether accompanied by AGN or not, are inherently self-regulating; the high-mass stars make it difficult for new generations of stars to form in the dense, central part of the galaxy where their winds and frequent SNe would disturb condensing clouds. A popular scenario on the subject suggests that as starbursts age, however, stellar winds decline and the production of interstellar dust declines. Because dust is blown away by photon pressure and the "corpuscular drag" from the stellar winds, the diminishing production of dust results in reduced extinction effects. Thus what once would have appeared as a starburst galaxy may later be seen as a LIRG, and still later as an ordinary spiral or elliptical galaxy with only modest reddening. In a study of the "dusty" radio galaxy MG1019+0535 (an apparent post-starburst galaxy), Manning & Spinrad (2001) attributed the mild self-absorption of its Lyα line to dust. The apparent coolness of the dust cloud, as constrained by sub-millimeter measurements, was attributed to the outward movement of dust clouds. Calculations placed most dust at a distance $\gtrsim$300 kpc from the galaxy, an estimated $\approx$ 500 Myr after the starburst that created it.

1.3 Merger Rates and Merger Signs

Even today, when typical large galaxy separations approach 1 Mpc, "collisions" of a minor nature are still common (on a dynamical time scale of $\gtrsim 10^8$ yr). Dynamical friction can then aid in slowing an encounter and accentuating the overlap in stars and gas we may term a merger. Our Milky Way galaxy is currently undergoing a few mergers (the Sagittarius Dwarf, and the two Magellanic Clouds).

Close examination of the nuclear regions of some spirals and E-systems indicate pluralities in their close nuclei – barely resolved by HST images. A dramatic new case is the "double-nucleus" of M31 (Statler et al., 1999). This "old" merger displays little indirect evidence of "action"; however, there are relatively nearby cases of on going mergers where the system gas is spewed out into obvious tidal tails. Examples are NGC 4038/9, the "antennae" galaxies (Levenson et al., 2000). NGC1569 (Stil & Israel, 1998) is a starburst in which a nearby H I companion has been implicated in stimulating the starburst.

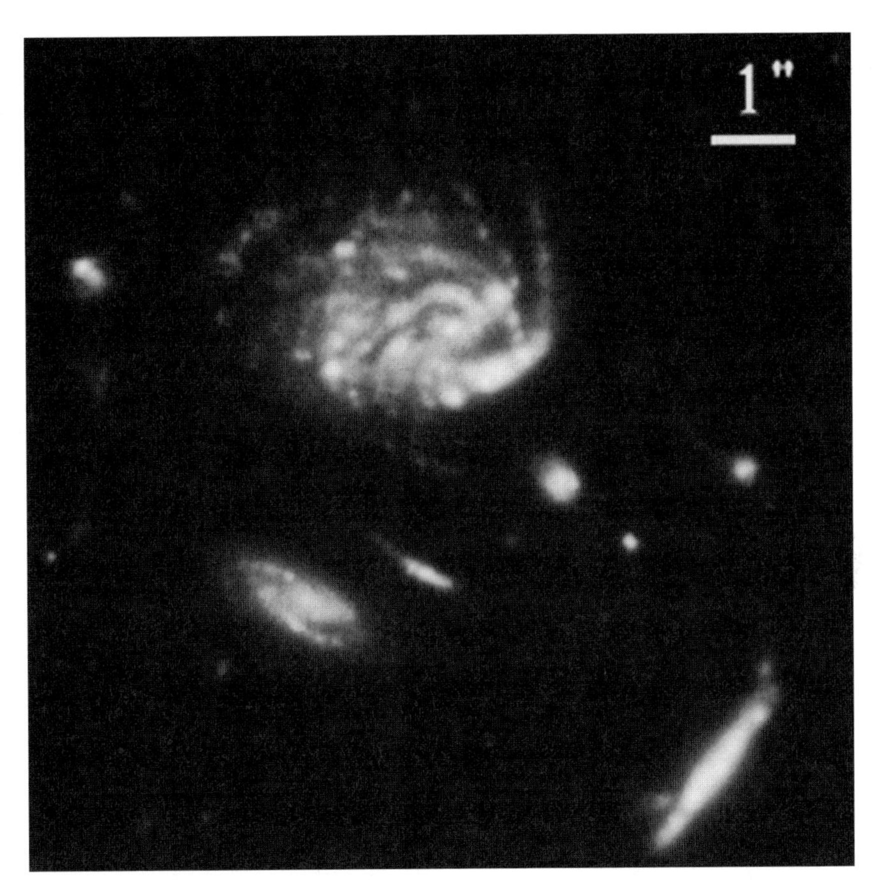

FIGURE 6: An enlargement of a portion of the Hubble UDF (a product of the HST and the Space Telescope Science Institute; STScI) showing a coalescing group of galaxies at $z \approx 1$; image courtesy of H. Yan and R. Windhorst. The face-on spiral seems perturbed and asymmetrical. The edge-on spiral appears to be in the process of cannibalizing two dwarf galaxies. The red galaxy appears to have the morphology of a spiral; the highlight on its lower arm suggests a burst of SF. Other smaller galaxies in this image have sizes of about $\sim 0.2''$, suggesting a physical size of about 1.6 kpc at $z = 1$. One must remember that most mergers are of unequal sized objects. (Reproduced in colour in the colour section.)

In the Hubble Ultra Deep Field (UDF), one can see galaxy groups at modest redshifts ($z \lesssim 1$) with a high incidence of disturbed morphologies, relative to the local Hubble tuning fork. Figures 6 and 7 show two small groups. The larger galaxies of the groups span about 3 arcsec, the angle subtended by a $\sim$25 h_{70}^{-1} kpc diameter galaxy at $z = 1$ under standard ("concordance") cosmological parameters (see Chapter 3). In addition to "imminent" mergers of comparable-sized galaxies, one can see imminent minor mergers, such as with the edge-on spiral in Fig. 7, which appears to have two dwarfs in the process of merger.

In a study of relatively nearby galaxies from the CNOC2 (Canadian Network for Observational Cosmology) survey of some 5000 nearby galaxies, Patton et al.

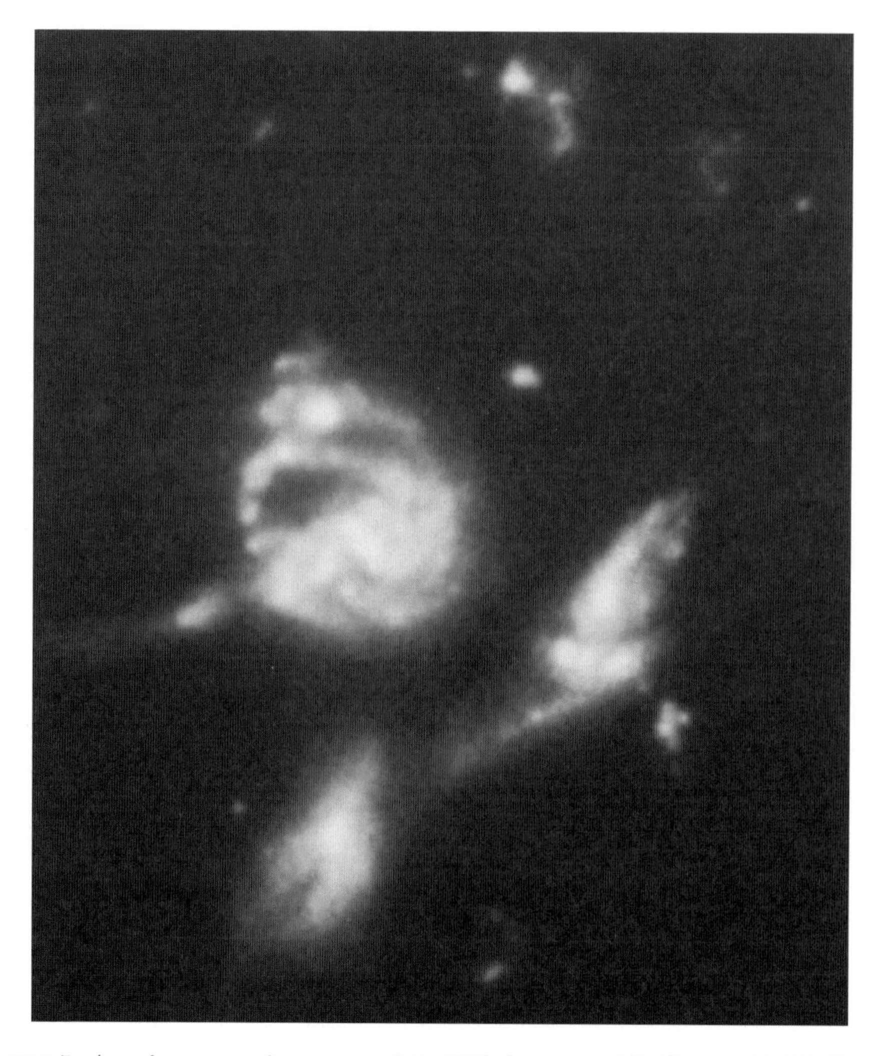

FIGURE 7: An enlargement of a portion of the UDF (courtesy of H. Yan and R. Windhorst) showing a coalescing group of galaxies at $z \approx 1$. Though it is difficult to be sure, each of the major galaxies appears disturbed, probably from interactions. Notice the large number of very small galaxies around this group – dwarf galaxies which are probably physically associated with the coalescing group. (Reproduced in colour in the colour section.)

(2002) find that ~3% of galaxies in the range $-21 \leq M_B \leq -18$ are close dynamical pairs ($5 < d < 10h^{-1}$ kpc). They infer an evolution rate of accretion $\propto (1+z)^{2.3\pm0.9}$. Thus 15% of present epoch galaxies in this range of luminosity have undergone a major merger since $z = 1$.

Signs of modest mergers in E-galaxies in the past can be found in the "photometric shells" visible in the exteriors of model-subtracted images. These are likely stellar remnants of mergers in the past, perhaps 10^9 years ago.

Arp 220 may be a nearby example of an earlier stage of merging, with an extended structure reminiscent of tidal tails. Morphological analysis shows signs of multiple nuclei, and Chandra observations again find compact regions of X-ray emission that may be indicative of multiple black holes with total X-ray luminosity of $\sim 1.2 \times 10^{41}$ erg s^{-1}.

On a larger scale, we shall see that the high resolution of Chandra X-ray observations allows researchers to analyze details of mergers on the galaxy cluster scale (i.e., $\lesssim$1 Mpc).

1.4 The Nature of AGN

In the nearby Universe, radio galaxies and quasars are quite rare. Our own Milky Way center displays only a few of these exotic phenomena; we detect no radio jet; an optically luminous nucleus is also lacking in our Galaxy. Of course we do have a central $\sim 3.7 \times 10^6$ $M_\odot$ black hole (BH).

In a minority of luminous nearby galaxies we note substantial contributions of non-stellar or even non-thermal radiation, in addition to the normal components of starlight and emissions from the gas of their ISM. In this sub-section, we introduce Seyfert and radio-loud galaxies. Seyferts, whose host-galaxy is a spiral, and radio-loud galaxies, whose host is almost always an E-galaxy, are AGN whose luminosity is probably driven by the same physical factors, gravitational potential energy of the nuclear condensation amplified on small scales by a massive black hole (MBH).

1.4.1 Seyfert Galaxies

Seyfert galaxies are luminous spirals with active nuclei, but they are observed to be of two general types, types 1 and 2. Type 1 produces broad emission lines, while type 2 produces relatively narrow lines. In the "unified" model of Seyferts, presented in Fig. 8, a black hole is surrounded by a dusty torus, which obscures the visible light from most vantage points (making it a Seyfert 2), but being seen as a bare AGN from other advantageous positions (a Seyfert 1). The "broad-line region" (BLR) is sampling the huge radiation field and turbulence around the AGN, and emission lines can be expected to have a large dispersion. Some of the width is due to rotation of the BLR. Narrow lines have been attributed to clouds at modest galactocentric radii that are illuminated by the bare nucleus. Their larger distances mean cloud velocities are smaller, hence lines are narrow.

Star-formation in Seyfert galaxies takes place in two general areas: in the disk, usually concentrated in a ring some distance from the nucleus, and in the torus, close to the nucleus. It is interesting to note that Seyfert 2 galaxies are often found to have young ($\approx 10^6$ yr) nuclear starbursts (e.g., Cid-Fernandes et al., 2001), as revealed by HST observations of C IV and Si IV P-Cygni lines. It has been suggested that the cause of the nuclear starbursts is that when the AGN is active the black hole accretion disk is loaded with gas, and this can satisfy the condition for gravitational instabilities that may lead to nuclear star clusters.

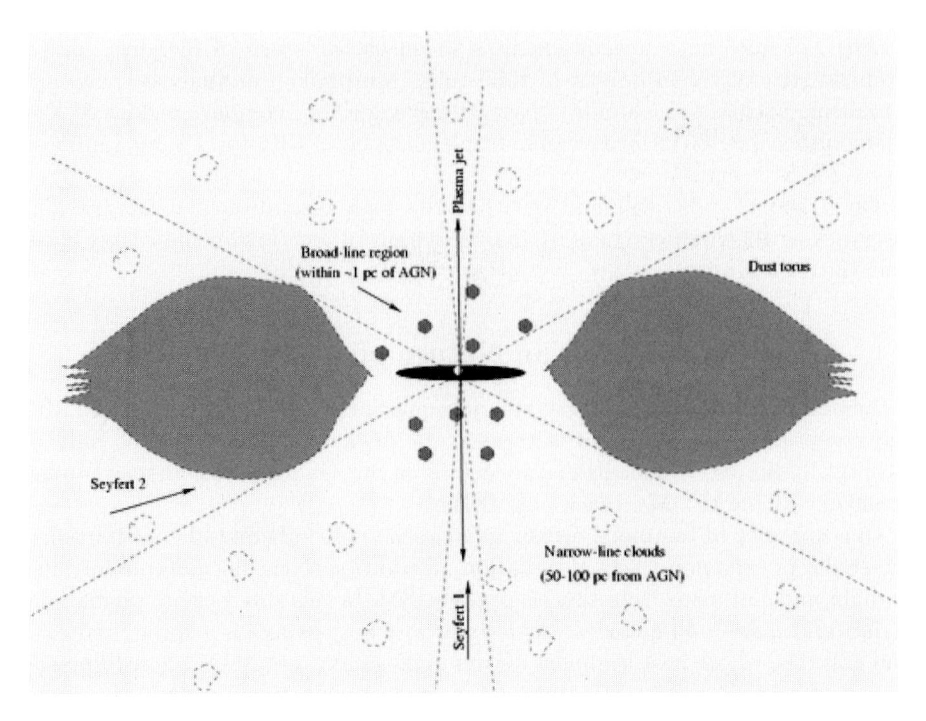

FIGURE 8: The unification model for AGN. Inner ellipse is the accretion disk (white dot represents the AGN itself), grey blobs represents the dust torus, shaded small clouds represent broad-line clouds, within about 1 pc of the nucleus, and the white clouds represent the narrow-line clouds, which lie from 50 to 100 pc from the nucleus. Viewed within the wider angle (top and bottom), the AGN is viewed as a (broad-line) Seyfert 1, but if the dust torus obscures the nucleus, the narrow-line clouds are most prominent in AGN spectra.

At higher redshift, we are unable to see the host galaxy, and AGN appear as quasi-stellar. However, while more quasars are elliptical, host galaxies for QSOs may be spirals or elliptical, or even host galaxies that are morphologically "messy".

1.4.2 Nearby Powerful Radio Galaxies

The nearest moderately strong radio galaxies are Per A (NGC 1275) and Virgo A= $M87$ (NGC 4486), with distances $\gtrsim$ 15 Mpc. In a volume with a 15 Mpc radius we aim to get the local space density of "strong" radio sources – within $z = 0.00375$ or so. Given that the local Schechter function normalization is $\phi^* = 1.65 \times 10^{-2} h^3\,\mathrm{Mpc}^{-3}$ (2dF result) , we expect $74h^3$ galaxies with $L \geq L^*$ within 15 Mpc. So the radio galaxies are moderately rare.

Despite their overall rarity, radio galaxies inhabit an enlarged fraction of the most luminous E galaxies. Strong radio sources are almost never hosted by spirals. In fact, the optical identification of powerful radio galaxies has proven a useful method to locate gE systems. This is apparently true both nearby and at large redshifts.

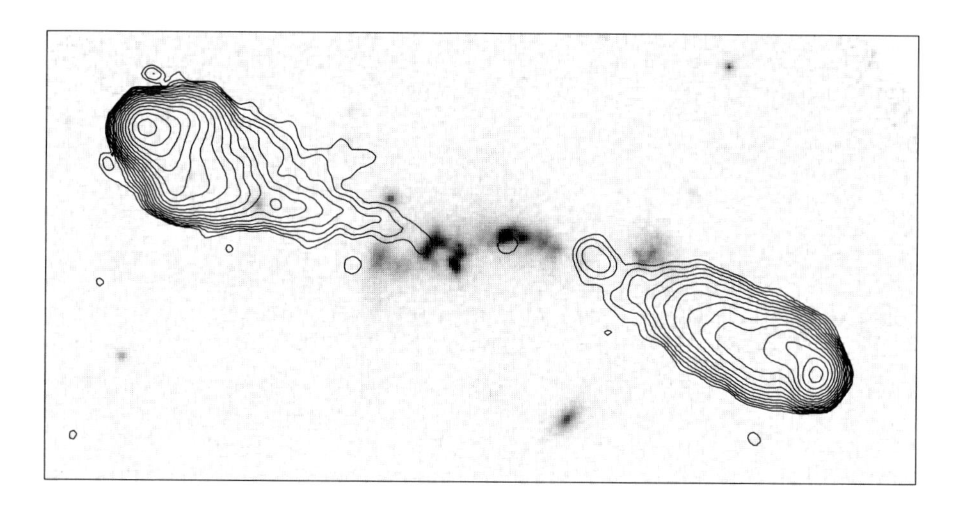

FIGURE 9: HST Planetary Camera image of the radio galaxy 3C 324 imaged with the F702W filter of the HST planetary camera (courtesy of Mark Dickinson). The radio contour overlay shows the 5 GHz (~6 cm) VLA map of this powerful radio galaxy (Fernini et al., 1993).

The morphology of radio galaxies points to some facets of their interesting and unusual nature. The radio emission itself is morphologically atypical; the strong sources have images in the form of lobes of synchrotron radiation. We illustrate, in Fig. 9, a "Cyg A" type morphology – actually the distant 3C 324 system ($z = 1.206$; Dickinson 1995). There is a preferred axis of activity; we theorize that the massive BH nucleus is surrounded by a rotating accretion disk, and the "radio jets" escape perpendicularly to the accretion disk along the rotational axis. Our concept features a very small inner-disk radius. Unfortunately, this inner axis is not necessarily purely along the minor axis of the (slowly) rotating E galaxy, as seen on kiloparsec scales.

QSO's or quasi-stellar-objects, have the rare and luminous active nuclei whose profuse radiation at many (broad-band) wavelengths has drawn the attention of astronomers from the 1960s to the present time. The radio-loud versions were the first to be optically identified, and their critical luminosities uncovered. These QSOs, called quasars (QSRs) with the "final R" indicative of their radio emission, were a complete mystery until Greenstein & Schmidt (1964) recognized the optical emission line pattern of two bright QSRs and their spectra indicated their surprisingly large redshifts. This was in 1963!

Since the 1960s, our ability to successfully catalogue a large number of QSOs has increased; modern QSO surveys now measure $\sim$ 10 to 20/□° (where □° is read, "square degrees"). The Sloan Digital Sky Survey (SDSS) and the 2dF Australian survey will find (eventually) a total of about 20,000 (the QSO surface density could amount to $\sim 40/\square^\circ$ to $m_B = 21$). We discuss these landmark surveys in more detail in §1.7.

Many of these are at "modest" redshifts, say $\overline{z}_Q \lesssim 2$, but they have a large range. More gradually, our ability to physically understand QSOs has also developed; a strong consensus now proposes a common scenario for AGN – feeding a "monster" black hole, with $M_{\rm BH} \geq 10^7$ M$_\odot$, through an accretion disk of $\sim$ solar system size. To actually produce a substantial inflow so close to the host galaxy nucleus (critical to its feeding) may require some non-symmetric inflow caused by a galaxy merger. Probably other massive BHs exist without central "dumping" at a large rate – they are likely to be relatively quiet and/or lower mass nuclei, such as the BH "observed" indirectly in our Milky Way – far from the level of QSO luminosity/activity.

From the surrounding stellar or gaseous kinematic behavior (like a mini-rotation curve) one can estimate the BH mass; for our (inactive) MW BH, $M_{BH} \sim 3.7 \times 10^6$ M$_\odot$ (Ghez et al., 2003; Genzel et al., 2003), while for active Seyfert galaxies (spiral hosts of mini-QSOs $M_{BH} \sim 10^7$ M$_\odot$ (Onken et al., 2003), and for QSOs their BH masses, less certainly, are about 10^9 M$_\odot$ (with a considerable dispersion). Now we can estimate this mass and the QSO luminosity in a simple calculation equating their gravity to the outward "push" of radiation pressure – this is called the Eddington luminosity; for a spherical mass in balance,

$$L \leq L_E = \frac{4\pi\, c\, G\, m_H}{\sigma_T}, \tag{9}$$

where σ_T is the Thomson cross-section for scattering free electrons. In astronomical observers' units, this translates to

$$L_E = 1.3 \times 10^{38} \left(\frac{M}{\mathrm{M}_\odot}\right) \mathrm{erg\,s^{-1}}, \tag{10}$$

(Osterbrock, 1989). Of course, in any specific case of a QSO or Seyfert nucleus, the Eddington luminosity (integrated) suggested above may only become an upper bound. The BH conversion of gravitational potential energy to radiation may then lead to a fueling time scale; modern values of fuel exhaustion are around 10^8 yr for QSOs – relatively short intervals compared to our assumed host-galaxy ages.

1.4.3 X-ray Bright Nuclei

X-ray bright nuclei are now being realized as an important contributor to non-stellar emission in normal-appearing galaxies.

As we have seen, it is often the case that the BH nuclei of galaxies are hidden by dust – often the result of associated nuclear starbursts. Though extinction generally is more effective against higher-frequency radiation, for very high-frequency radiation, the cross-section for absorption declines faster than the number of potential absorbers. Thus, it is found that while "soft" X-rays (i.e., lower energy; 0.3

to 3 keV) are effectively absorbed by dust tori, while "hard" X-rays (3 to $\sim$ 8 keV, can penetrate the torus, and are thus detectable by the Chandra X-Ray Telescope. Chandra has good spectral sensitivity to hard X-rays. Energies of about 8 keV, corresponds to a temperature of about 8×10^7 K, such as is found in a rich cluster of galaxies, with a velocity dispersion ≈ 1000 km s^{-1}.

Chandra images often show many point sources – some associated with X-ray binaries of various sorts, and some are less variable, stronger sources that are likely BHs. The Chandra X-Ray Observatory has thus been utilized to search for bright, point-like X-ray sources as black hole candidates. For instance, the brightest of nine X-ray sources in the starburst galaxy M82 (Wan et al., 2002; McDonald et al., 2002) is thought be an X-ray binary with an intermediate mass black hole. This source coincides with the second-most luminous infrared feature, possibly indicative of a nuclear star cluster. The others could be smaller BHs.

1.5 Groups, Clusters, and Rich Clusters of Galaxies

The same factors that cause matter to agglomerate in galaxies cause galaxies to agglomerate in groups. A notable poor group is the Milky Way/Andromeda/M33 – i.e., the Local Group. Groups are typically dominated by spiral galaxies. As larger groups are considered, the main constituent galaxies shift from spirals to lenticulars and ellipticals. The intra-cluster medium plays a large part in erasing the normal appearance of spirals that fall within its grasp (i.e., star-forming regions; spiral arms). In addition, in the denser cluster medium, the frequency of galaxy collisions must be significantly greater. The latter fact may help to explain the occasional presence of super-massive cD galaxies in their midst. It may also explain the high hot (X-ray) gas to stellar mass ratio in clusters (about 2/3), and why the mass of the million-degree gas is observed to increase with cluster mass (Wu & Xue, 2002). For as the virial temperature of the halo increases, galaxies will experience greater ram-pressure stripping.

George Abell (Abell, 1958) first systematized the detection of rich clusters of galaxies as a way to investigate the large-scale distribution of matter. Clusters were cataloged according to seven distance groups, according to the apparent magnitude of the tenth brightest cluster member. Clusters are assigned to one of six richness classes according to the galaxy counts. His study included a total of 2712 clusters, from which a homogeneous sample of 1682 clusters was drawn, which is now known as the Abell Catalogue. From this, a very rough idea of the spatial and mass distribution of clusters can be gleaned.

Perhaps the two most-studied galaxy clusters are the relatively nearby Virgo and Coma clusters. Virgo is not an especially rich cluster, but is still in the process of assembly. It is a loose, or "open" cluster of galaxies. The closest truly rich cluster of galaxies is Coma, about 70 h^{-1} Mpc distant. The distribution of galaxies, and to some extent, the hot gas near its core, is quite symmetric. Many E-rich, or cD clusters have symmetric X-ray gas distributions; some show more complex distributions of 10^7 to 10^8 K gas (e.g., Maughan et al., 2003).

However, the observational challenges to accurately estimating the richness of distant clusters are many. These involve the need to separate foreground or background galaxies, and sometimes even superimposed clusters. These complications argue strongly for a better, more secure methodology for assessing cluster richness and mass.

Two main methods have been devised. The chronologically first method utilizes the effects of the virialization of cluster halos, and equipartition of energy. The result is that the kinetic energy per unit mass in the cluster is roughly equal to the depth of the potential well. This in turn determines the temperature of the gas. The temperatures in rich clusters approach 10^8 K, making the gas a source of X-ray photons. Thus, clusters can be detected by their X-ray flux. Since the emission rate per unit mass is proportional to the density, it is possible to model the distribution of mass in the cluster. Cluster gas density is best modeled as having a core turning to an inverse-square density profile. A classic example of this was presented in the study of M87 in the Virgo Cluster (Fabricant & Gorenstein, 1983), roughly 15 Mpc distant. This centrally positioned giant elliptical galaxy has a halo that merges with that of the cluster. Observed with the Einstein X-Ray Observatory, the virial temperature consistent with $kT \simeq 3$ keV, is $T_{\rm vir} \simeq 3 \times 10^7$ K. Modeling the gas as hydrostatically supported, Fabricant and Gorenstein were able to derive the mass distribution from the X-ray luminosity.

A series of X-ray telescopes have been put in orbit, culminating in the deployment of the Chandra X-Ray Observatory in 2001. It represents a great advance in spatial and spectral resolution. These advances enable astronomers to detect the X-ray luminosity of clusters at redshifts approaching unity, and begin to study their substructure. For instance, Chandra studies of the $z \sim 0.8$ cluster CL J0152.7-1357 (Maughan et al., 2003) find it is composed of a northern and southern subclusters, each massive and X-ray luminous, in the process of merging. Further analysis reveals excess emission between subclusters, suggestive of a shock front.

One may also estimate the cluster mass using the fact that the mass density of a cluster can act like a lens; bending light and magnifying images of background objects. This is the basis of the second method. It has two regimes; weak, and strong lensing. The former requires an intensive analytical investment to discern the slight distortional field imposed on background galaxies, but strong lensing can be seen directly. The former is used for estimating the large-scale distribution of mass in the cluster. With strong lensing, multiple images of background galaxies may be seen within the "Einstein radius", magnified by the cluster potential. These data can be used with lensing models to constrain the projected mass distribution in detail. Often researchers find the halo can be resolved into multiple sub-halos superimposed over a smooth "cluster" halo. These methods have been used as to substantiate the independent estimates of the cluster mass distribution based on the X-ray flux.

A third avenue for investigating clusters is the Sunyaev–Zeldovich (Sunyaev & Zeldovich, 1969) effect (SZE). This effect is a perturbation of the cosmic background radiation (CBR) by the interaction of CBR photons with plasmas, such

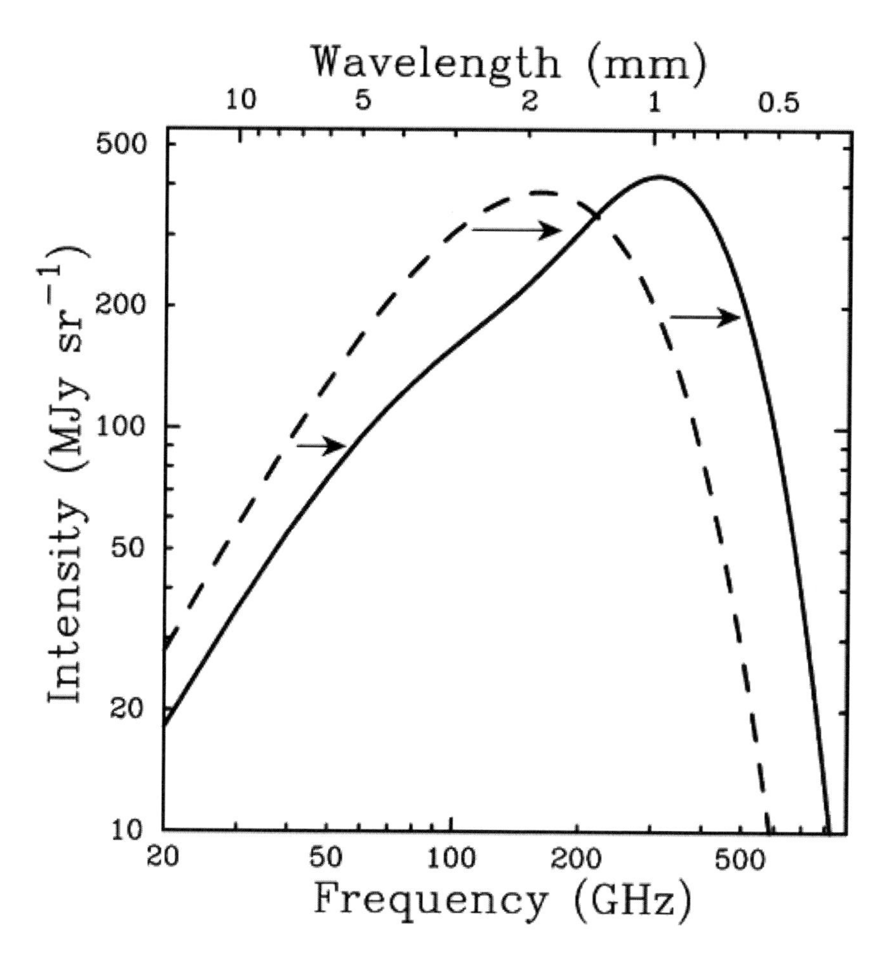

FIGURE 10: The SZ effect results when microwave background photons traverse a rich cluster of galaxies, and experience inverse Compton scattering from hot electrons in the inter-cluster medium. In this illustration, from Carlstrom et al. (2002), the dashed curve shows the original Planck spectrum, and the solid line shows a typical SZ result, with lower-energy photons being scattered to higher energies. The flux is given in Jy m^{-2}. Reprinted, with permission, from the Annual Review of Astronomy and Astrophysics, Volume 40 (c)2002 by Annual Reviews www.annualreviews.org.

as might be found in the dense, ionized IGM of galaxy clusters. These distortions of the smooth CBR provide another method of detecting clusters against the confusion of optical data (see Fig. 10).

These studies appear to suggest a fairly tight mass–X-ray temperature relation in galaxy clusters. The distribution function of cluster masses (or temperatures) in time interests astronomers because, in the picture of hierarchical assembly, the largest structures are expected to have formed most recently. But just how recently this occurs is a function of the density parameter Ω_m*, and other cosmological parameters. An early formation of clusters ($z \gtrsim 1$) would imply a low-density, perhaps "open" Universe. We discuss these issues in more detail in Chapter 5.

1.6 Astronomical Instrumentation at the Millennium

As an interlude to our concentration on galaxies, we conclude this initial chapter with a discussion of some aspects of modern astronomical instrumentation that allow us to successfully observe galaxies both near and distant.

1.6.1 Earth-based Optical Telescopes and their Auxiliary Equipment

The instruments currently in use, or which exist as firm plans for construction, are both ground-based and space-based facilities. At this time, astronomers are just beginning to routinely utilize the power of new very large optical-IR telescopes. Thanks to improved mirror technologies (for both segmented primaries and monolithic mirrors) several collaborations have recently constructed much larger telescopes than the "giant of the past", the Hale 5-m (200″) reflector. See Table 7 for a list of 6-m to 10-m telescopes.

Even more relevant to this book are the noticeable improvements in detectors, particularly those leading to the wide adoption of charge-coupled devices (CCDs) for many Astronomical tasks in the extended optical window. Much of our current observing on nearby and distant galaxies is now undertaken with CCDs and/or arrays of CCDs.

Our recent successes in using CCDs is a result of three of the device's qualities. First, they are high-quantum-efficiency detectors (varying over wavelength, but typically $QE \geq 0.6$ over much of the optical window). Secondly, they are usually very linear detectors. Third, they are, of course, imaging detectors with many pixels, and thus are very amenable to low-contrast spatial imaging and processing. The ease of manipulating digital data has sparked the observational computer-processing era in our field.

CCD surface photometry has been important in defining the photometric properties of relatively nearby galaxies discussed earlier in this chapter. A wide area of advantage is in defining the sky "foreground" without oversubtracting the outer low-surface-brightness extensions of galaxy light. In almost every consideration the imaging digital detector has improved our photometric abilities over those possible with "aperture integration", e.g., photomultipliers or small arrays of silicondiodes.

The structural character of CCDs is that they are integrating photon counting (one (emergent) electron for each detected photon) devices. Their typical sizes used to be $\simeq 1\,\mathrm{cm}^2$, with 500×500 pixels, but in recent years the technology has made available CCD "chips" of dimension about 2K × 2K (or even 4K × 4K in a mosaic combination.

---

*Ω_m is the matter density in units of the critical density, given by the equation, $\rho_\mathrm{crit} = \frac{3H_0^2}{8\pi G}$, where G is the gravitational constant and H_0 is the Hubble constant.

TABLE 7: The World's Largest Optical Telescopes

Aperture (meters)	Name	Location	Altitude (meters)	Comments
10.0	KeckI KeckII	Mauna Kea, Hawaii	4123	36 segmented mirrors: future optical interferometry
9.2	Hobby -Eberly	Mt. Fowlkes, Texas	2072	inexpensive,; segmented spectroscopy only
8.3	Subaru	Mauna Kea, Hawaii	4100	NAOJ; wide angles.
8.2	Antue Kueyen Melipal Vepun	Cerro Paranal, Chile	2635	operate independently; future units of VLT
8.1	Gillett	Mauna Kea, Hawaii	4100	aka Gemini North
8.1	Gemini South	Cerro Pachon, Chile	2737	twin of Gemini North
6.5	Walter Baade	Las Campanas Obs. Chile	2282	aka Magellan I;
6.5	Landon Clay	Las Campanas Obs, Chile	2282	aka Magellan II
6.0	Bolshoi Teleskop Azimutalnyi	Nizhny Arkhyz, Russia	2070	aka Large Altazimuth Telescope

The CCD data comes to the computer interface based upon internal chip charge transfer and collection. The efficiency of the charge-transfer is important, and in modern CCD devices it is close to unity (perfect) and, in any case, modelable. Several useful technical articles on CCDs have appeared and are useful to Astronomers. An interested reader may wish to sample one even older reference (Djorgovski, 1984) and several in Robinson (1988).

In the ultraviolet ($\lambda < 3000$ Å), most CCDs have a lowered quantum efficiency. Then the detectors of choice are photon-counting devices, photocathodes of CsI and Cs_2Te, especially efficient in the UV below 3000 Å (recently in active use in the Space Telescope Imaging Spectrograph (STIS)).

On the other side of the visible, new near-IR detectors and cryogenic cameras have helped in the development of new IR instruments for ground-based and space use.

Both ground-based and space-based plans for current and future instruments place some emphasis on new IR techniques; lower backgrounds/foregrounds, and modern detectors will help to achieve some considerable improvement in IR imaging and spectroscopy of galaxies.

The IR instruments available on the new generation of 8-m class telescopes can now be used for imaging in the isolated mid-IR "windows" in the atmospheric absorption with small fields of view. Obviously this can help in studies of dusty galaxy environments. These use Si:As detector surfaces. Also spectroscopy at shorter wavelengths with InSb detectors (0.8–5.5 μm) is now practicable to fairly faint levels, although the night sky (OH, in particular) adds a particularly strong foreground emission hurdle to such research. For brighter objects, resolutions of $R = 2000$ are quite practical. Thus, spectral studies of extragalactic star clusters in the J, H, and K bands have become practicable.

Some of the same detectors, behind different imagers and spectrometers, may be flown as important instruments upon the next generation optical space telescope (see §1.5).

In general the IR is a useful spectral domain for galaxy observations in cases of dust-reddening, or in situations where the redshift has moved important spectral diagnostics to longer wavelengths. That will be of import in later chapters.

A modern efficiency-aid is the multi-object spectrograph or spectrometer. This gives the observer of small galaxies and stars a large "multiplex advantage", leading to simultaneous exposures on many targets, with $\approx$25–100 slitlets covering an area of perhaps $\sim 1/6\ \square'$ per slit mask. The slitlets in the mask must be manufactured in advance. Another option is to use fiber-optical positioning and hence feeding to a fixed spectrograph by the lengthy fibers. The Australian 2dF (two-degree field) and 6dF (6°) spectrograph surveys of galaxies and QSO's illustrate this technique. The 2dF spectroscope is outfitted with 400 robotically placed optical fibers. The entire survey covers $\sim 2000\ \square°$to a mean redshift of ~ 0.1 (e.g., Taylor et al., 1997; Schilling, 2003; Norberg et al., 2002). The 6dF, with 150 robotically placed fibers, will cover the southern hemisphere, but to a shallower depth (e.g., Watson et al., 2000). Optical fibers yield excellent spectra for star-like (small angular extent) objects; slits are more useful for extended objects, such as low-z galaxies.

1.6.2 Radio Telescopes

Radio telescopes are an important window into the Universe since the atmosphere is transparent to these long-wavelength photons. However, since they are long in wavelength, the diffraction limitations mandate large telescopes. Currently there are two essentially different design approaches in radio astronomy. One uses large single antenna, and the other uses arrays of telescopes in order to synthesize a larger baseline, enabling effectively larger apertures with respect to resolution. The angular resolution of a single telescope of diameter d observing with light of wavelength λ is, of course,

$$\theta = 1.22\ \frac{\lambda}{d}\ \text{radians.} \qquad (11)$$

With a synthesized aperture, very high resolution can be attained (from VLA to VLBA; see Table 8). Another particular use of arrays is in mapping; the number of "pixels" produced is n^2, where n is the number of antenna elements, while the field of view over which the pixels are arrayed is that of a single element of the array (see Eq. 11). Single-dish telescopes must use separate moves in order to perform mapping. However, rough mapping can be accomplished when multiple detectors are used at the focal plane. Basic information on the largest single dish and interferometric detectors are presented in Table 8.

1.6.3 The Spitzer Space Telescope

The idea which gave birth to the Space InfraRed Telescope Facility (SIRTF), now renamed after Lyman Spitzer, was inspired by the need to extend the reach, resolution, and accuracy of the Infrared Space Observatory (ISO). The Spitzer Space Telescope (SST) is a joint project of NASA and the ESA, which provided the first good look at the universe in the mid- to far-infrared (Werner et al., 2004). It has a wavelength sensitivity ranging from ~2.5 to ~ 180 μm, though beyond 30 μm it becomes less sensitive. Spitzer has a range of sensitivity very similar to ISO, with imaging and spectroscopy in the range of 3.6 to 200 μm, but has significantly better spatial resolution, especially from 3.6 μm to ~20 μm. This telescope is in an Earth-trailing orbit, which will cause it to gradually drift away at a rate of ~0.1 AU per year. Its cryogenic supply is expected to last for 2.5 to 5 years. Among its "legacy" observing programs is a study of the Galactic mid-plane, a deep survey of galaxies to help understand galaxy formation over a large redshift range, a study of proto-planetary disks around young stars, and a study of the star-forming ISM of nearby galaxies.

1.6.4 Telescopes of the Future

State-of-the-art astronomical instrumentation generally takes about 20 years from the concept to the product. There are a number of "next-generation" telescopes in the process of being built, including giant optical telescopes (the OverWhelmingly Large (OWL) telescope is 100 meters diameter); the Square Kilometer Array would have tens of thousands of individual cm-wave detectors spread over thousands of miles; Atacama Large Millimeter Array (ALMA) will be the next-generation millimeter array. In addition, the next generation space telescope, now called James Webb Space Telescope (JWST) is well into the planning stage, and is scheduled to be launched in 2011. This telescope will have a light-gathering power about 10 times that of the Hubble Space Telescope (HST), and its infrared sensitivity will enable it to detect highly redshifted galaxies. It will also be useful in the detection and study of extrasolar planets.

We will discuss these and other new observatories in Chapter 6.

TABLE 8: List of Modern New and Large Radio Telescopes

Single dish				
Name	Location	Wavelengths	Size (m)	Comments
Bonn	Effelsberg, Germany	m/cm	100	not new
GBT	Greenbank, W. Virginia	cm	100×110	actuator-controlled shape
Arecibo	Puerto Rico	m/cm	305	newly upgraded
Jodrell Bank	Cheshire, England	m/cm	76	
Parkes	Australia	m/cm	64	being upgraded
Nobeyama	Japan	cm/mm	45	-
Interferometers				
Name	Location	Wavelengths	Size	Comments
GMRT	Narayangaon, India	m/cm	30×45 m	Seeking H I emission at hig-z
VLA	New Mexico, USA	m/cm	27×25 m	"Y-shaped" array
WSRT	Netherlands	cm	14×18 m	linear array
ATCA	Australia	cm	6×22 m	linear array
VLBA	scattered sites, USA	cm/mm	25 m each	longest baseline 8600 km
OVRO	Owens Valley, CA, USA	mm	6×10.4 m	moving to CARMA
BIMA	Hat Creek, CA, USA	mm	9×6 m	moving to CARMA
CARMA	Owens Valley, CA, USA	mm	various	OVRO + BIMA+ 8×3.5-m dishes
ATA	Hat Creek, CA, USA	cm/mm	350×6 m	completion 2006
NRT	Nobeyama, Japan	mm	6×10 m	linear array
ALMA	Atacama Desert, Chile	mm	64×12 m	completion 2011

1.7 The Role of Large Sky Surveys

Galaxy redshift surveys are not new to astronomy, and would seem to add little more to our knowledge other than some more redshifts and statistical information about galaxies. However, in the closing years of the 20th century, two major surveys were undertaken, the 2dF galaxy redshift survey (where 2dF stands for 2-degree field; 2dFGRS), and the Sloan Digital Sky Survey (SDSS, or simply Sloan). The 2dF uses the 3.9-meter Anglo-Australian Schmidt telescope, while the Sloan survey uses a specialy built 2.5-meter telescope in New Mexico. We also include in this sketchy account the DEEP/2 survey, which makes up in depth what it lacks in sky coverage.

While the 2dF is exclusively spectroscopic, the Sloan survey is primarily an imaging survey in five photometric bands, with follow-up spectroscopy. The Sloan uses a drift-scan technique for imaging so that a given galaxy will be recorded sequentially in each of the spectral bands. The 2dF system can take spectra of up to 400 galaxies per observation. The SDSS system, which can place fibres closer together, can measure 640 relatively "nearby" galaxy spectra at a time. While the 2dF covers about 5% of the sky, the Sloan survey will observe about 25%; essentially the low-galactic extinction of the northern galactic caps with only slight coverage in the southern sky. The Sloan survey will eventually observe $\sim 5 \times 10^7$ galaxies, with spectroscopic observations of perhaps 10^6 galaxies. The median redshift in these surveys is around $z = 0.1$. The smaller 2dF has taken spectra of 221,283 galaxies out to a redshift $z = 0.3$, distributing its coverage roughly comparably between northern and southern skies.

The Sloan survey can use galaxy colors to sort galaxies into groups. They find two major populations; a redder population comprised of elliptical and S0 galaxies (systems that have de Vaucouleurs $r^{0.25}$ profiles), and a bluer population with roughly exponential surface brightness profiles (Blanton et al., 2003; Hogg et al., 2002). While red galaxies constitute one-fifth of the population, they produce two-fifths of the light (Hogg et al., 2002). The 2dF can analyze sub-populations of galaxies using spectral types. For a more thorough review of these large surveys see Fairall (2004)

At deeper levels of analysis, these surveys are being used to study redshift space distortions (Peacock et al. 2001 for the 2dF, and Zehavi et al. 2002 with limited Sloan data), and the spatial correlation functions and power spectra (Percival et al. 2001 with the 2dF data). These products may be used to extract cosmological parameters. The utility of this approach is that if low-redshift clustering scales are identified as caused by the anisotropies such as those seen in the Cosmic Microwave Background (CMB) observations (Spergel et al., 2003), the two distributions may constrain cosmological models that evolve from the recombination epoch to the present. This identification is made under the assumption that the perturbations of the CMB represent the weakly clustered mass distribution at the period of decoupling which evolved to that which is observed today. Thus, numerical simulations consistent with the CMB findings can be extended to the current epoch, and the quality of the agreement of model and observations can be assessed. In this way, 2dF and SDSS data can be used as a consistency check, or to refine the analysis of new CMB-only results.

A recent addition to the large surveys is the Deep Extragalactic Evolutionary Probe (DEEP), a coordinated survey of galaxies in selected fields in the redshift range $0.7 \lesssim z \lesssim 1.4$. The project is headed by astronomers from U. C. Berkeley (M. Davis and colleagues), U. C. Santa Cruz (S. Faber and colleagues), and Caltech (C. Steidel, R. Ellis and colleagues). Its goal is to understand how galaxies formed, and characterize the evolution of large-scale structure. An underlying goal is to supply measures of the Universe which can be compared with cosmological numerical simulations, and the independent data from the analysis of the CMB. The survey plans to achieve these goals by looking at density distributions of the various galaxy types, their velocity dispersions, line-widths, and spectra at a mean redshift $\langle z \rangle \approx 1$. In a series of 1-hour exposures, the properties of 50,000 galaxies are to be measured. Three-hour exposures will be used to study systems some 1.5 mags below L^*. Cosmological parameters will be probed by studies of large-scale structure.

The DEEP survey is a two-phased project using the Keck telescopes to study the properties and distribution of $z \sim 1$ galaxies. Phase 1 used the LRIS spectrograph to study a sample of galaxies at a median redshift of about unity. Phase 2 of the DEEP project started with the installation of an advanced second-generation spectrograph on the Keck II telescope, DEIMOS, which increases the throughput for faint-object spectrography by a factor of 7. The survey is designed to have the fidelity of local redshift surveys such as the LCRS survey, and to be complementary with ongoing large redshift surveys such as the SDSS project and the 2dF survey.

The DEIMOS/DEEP or DEEP2 survey will be executed with resolution R=4000, so that linewidths and rotation curves can be measured for a substantial fraction of the target galaxies. DEEP2 will thus also be complementary to the VLT/VIRMOS project, which will survey more galaxies in a larger region of the sky, but with much lower spectral resolution and with fewer objects at high redshift.

1.8 Summary

Our first chapter makes an assessment of extragalactic objects in the low-redshift universe. Nearby galaxies, though older than galaxies seen at elevated redshift, provide a good first approximation of what might exist at high z. Locally, galaxies are morphologically well-characterized by the Hubble tuning fork diagram . Because groups form on time scales of gigayears, the morphology–density relation suggests that at least some Es are the result of mergers of spirals. Mergers are thus probably responsible for a lot of the enhanced E fraction in clusters (by $\sim 3\times$), while the enhanced S0 fraction in clusters (a factor of ~ 2) is plausibly due to ram-pressure stripping of gas from original spirals.

While spirals have disks whose surface brightness declines exponentially, Es, and the surface-brightness distributions of bulges of Sa and Sb galaxies have the

$r^{1/4}$ power law. Colors of E galaxies are red, and spirals are generally blue. Irregulars are bluer still. The specific gas-content as a function of type changes continuously across the Hubble sequence, so Es have a small $M_{\rm HI}/M_{\rm tot}$, and the bluer spiral galaxies have a large ratio, as seems reasonable.

At moderate distances, merger rates are thought to increase strongly with increased redshift. Mergers may drive SF, and accretion by MBH. MBHs are thought to be the engine of the AGN. Optical AGN tend to be Seyfert galaxies, which come in two types; unobscured (Type 1; broad-line) and obscured (Type 2; narrow-line). Radio galaxies are powered by MBHs as well, which produce jets that produce powerful radio "lobes". The physics of accretion appears to be limited by the Eddington luminosity, which is the luminosity needed to give a radiation pressure equal to the gravitational acceleration. The nuclei of MBHs are X-ray bright when actively accreting – emitted by the viscously heated accretion disk.

Groups and clusters of galaxies are the natural signposts of large-scale structure in the Universe. Their morphology is various, varying from spherically symmetric, dominated by a central cD galaxy or a subset of large ellipticals, to multi-modal structures obviously out of equilibrium. For rich clusters, the X-ray luminosity and surface brightness of clusters, together with weak and strong lensing analyses, provide complementary data to determine the degree of dynamical equilibrium, and mass of the clusters.

In the preceding quick assessment of earth- and space-based instruments, we have seen that there are many opportunities to observe the same object with different instruments, allowing astronomers to more tightly constrain the physical processes taking place in the object. We shall find that, when inspecting more distant objects, a deeper understanding of more local objects will often put us in a good position to better judge the nature of high-redshift objects.

REFERENCES

Abell, G. O. 1958, ApJS, 3, 211

Abraham, R. G., et al., 1996, ApJS, 107, 1

Adelberger, K. L., et al., 2003, ApJ, 584, 45

Bershady, M. A., et al., 2000, AJ, 119, 2645

Binggeli, B. 1987, in *Nearby Normal Galaxies*,(New York:Springer-Verlag) ed. S. M. Faber, p. 195

Binney, J. & Merrifield, M. 1998 *Galactic Astronomy* (Princeton, New Jersey: Princeton University Press).

Blanton, M. R., et al., 2003, ApJ, 594, 186

Branch, D. 1987, ApJL, 320, L23

Carlstrom, J. E., et al., 2002, ARAA, 40, 643

Cid-Fernandes, R., et al., T. 2001, in Revista Mexicana de Astronomia y Astrofisica Conference Series, 133

Coleman, G. D., et al., 1980, ApJS, 43, 393

Conselice, C. J., et al., 2003, AJ 125, 66

De Propris, R., et al., 2003, MNRAS, in press, astro-ph/0212562

de Vaucouleurs, G. 1959, in *Handbuch der Physik*, vol. 53 (Berlin: Springer-Verlag), 275

de Vaucouleurs, G. 1961, ApJS, 5, 233

de Vaucouleurs, G. 1963, ApJS, 8, 31

Dickinson, M. 1995, in *Fresh Views of Elliptical Galaxies*, ed. A. Buzzoni, A. Renzini, & A. Serrano, Vol. 86 (San Francisco: ASP Conference Series), 283

Djorgovski, S. 1984, in *Improvements to Photometry*, 152

Dressler, A. 1980, ApJ, 236, 351

Fabricant, D. & Gorenstein, P. 1983, ApJ, 267, 535

Fairall, A. P. 2004 in *Astrophysics Update*, ed., J. W. Mason, (Heidelberg: Springer-Praxis) p. 211

Feast, M. W. & Walker, A. R. 1987, ARAA, 25, 345

Fernini, I., et al., 1993, AJ, 105, 1690

Freeman, K. C. 1970, ApJ, 160, 811

Genzel, R., et al., 2003, Nature, 425, 934

Ghez, A. M., et al., 2003, ApJL, 586, L127

Greenstein, J. L. & Schmidt, M. 1964, ApJ, 140, 1

Hogg, D. W., et al., 2002, Astron.J., 124, 646

Kennicutt, R. C. 1992, ApJ, 388, 310

Kewley, L. J. & Dopita, M. A. 2003, in Active Galactic Nuclei: from Central Engine to Host Galaxy, meeting held in Meudon, France, July 23-27, 2002, Eds.: S. Collin, F. Combes and I. Shlosman. ASP (Astronomical Society of the Pacific), Conference Series, Vol. 290, p. 519., 519

Levenson, N. A., et al., 2000, ApJL, 533, L53

Longair, M. S. 1998, *Galaxy Formation* (Berlin: Springer)

Manning, C. & Spinrad, H. 2001, AJ, 122, 113

Marzke, R. O., et al., 1998, ApJ, 503, 617

Maughan, B. J., et al., 2003, ApJ, 587, 589

McDonald, A. R., et al., 2002, MNRAS, 334, 912

Morgan, W. W. 1958, PASP, 70, 364

Norberg, P., et al., 2002, MNRAS, 336, 907

Onken, C. A., et al., 2003, ApJ, 585, 121

Osterbrock, D. 1989, *The Astrophysics of Gaseous Nebulae and Active Galactic Nuclei*, p. 347. (Mill Valley, California: Univ. Science Books).

Patton, D. R., et al., 2002, ApJ, 565, 208

Peacock, J. A., et al., 2001, Nature, 410, 169

Pedlar, A., et al., 2003, in Revista Mexicana de Astronomia y Astrofisica Conference Series, 303–307

Percival, W. J., et al., 2001, MNRAS, 327, 1297

Roberts, M. S. & Haynes, M. P. 1994, ARAA, 32, 115

Robinson, L. B. 1988, *Instrumentation for Ground-Based Optical Astronomy: Present and Future* (New York: Springer-Verlag)

Sandage, A. 1961, *The Hubble Atlas of Galaxies* (Washington D.C.: Carnegie Institution of Washington)

Sandage, A. & Bedke, J. 1975, *Atlas of Galaxies* (Washington D.C.: NASA)

Schechter, P. 1976, ApJ, 203, 297

Schilling, G. 2003, S&T, 105, 32

Schreiber, N. M. F., et al., 2003, ApJ, 599, 193

Silk, J. 1999, PASP, 111, 258

Spinrad, H. & Peimbert M. 1975, *Stars and Stellar Systems*, Vol. 9, "Galaxies and the Universe", p. 37, eds. A. Sandage, M. Sandage, and J. Kristian

Spergel, D. N., et al., 2003, ApJS, 148, 175

Statler, T. S., et al., 1999, AJ, 117, 894

Stil, J. M. & Israel, F. P. 1998, AAP, 337, 64

Sunyaev, R. A. & Zeldovich, Y. B. 1969, Nature, 223, 721

Taylor, K., et al., 1997, in Proc. SPIE Vol. 2871, p. 145-149, *Optical Telescopes of Today and Tomorrow*, Arne L. Ardeberg; Ed., 145–149

van den Bergh, S. 1960a, ApJ, 131, 558

van den Bergh, S. 1960b, ApJ, 131, 215

Walker, A. R. 1992, AJ, 104, 1395

Wan, E., et al., 2002, American Astronomical Society Meeting, 201, 0

Watson, F. G., et al., 2000, in Proc. SPIE Vol. 4008, p. 123-128, *Optical and IR Telescope Instrumentation and Detectors*, Masanori Iye; Alan F. Moorwood; Eds., 123–128

Werner, M. W., et al., 2004, ApJS, 154, 1

Wu, X. & Xue, Y. 2002, ApJL, 572, L19

Young, J. S. & Scoville, N. Z. 1991, ARAA, 29, 581

Zehavi, I., et al., 2002, ApJ, 571, 172

Chapter 2

Which Properties of Galaxies can Likely Evolve (and be Measured)?

In this chapter we begin to evaluate selected properties of galaxies that are likely to systematically change over cosmic epochs. We'd like to anticipate some changes – can high-redshift galaxies evolve into the systems we observe nearby at low-z? And can galaxy morphologies change?

The most systematic of these potential evolutionary changes to galaxies are the large-scale effects of mutual gravitation, and the long timescales that may be involved.

The need to quantitatively describe the evolution of objects in a Universe that expands with time brings us into confrontation with ambiguities of measure that do not occur with local observations. For cosmological observations we initially view the objects (galaxies, clusters) as being at rest in a universe expanding at a rate defined by the assumed cosmology. Since the space between clusters is expanding, the wavelengths of light they emit are also expanding, and shifted toward the red. This expansion is parameterized by a scale factor, $R(t)$, or $R(z)$ which expands as the Universe as a whole.

Imagine that an astronomer wants to calculate the distance to a distant object. There are ambiguities associated with the distance when the objects are at cosmological distances. A *proper* distance is the spatial displacement between two points at the *same* cosmological epoch and is determined by the light-times for travel between points. For cosmological displacements, this becomes inaccurate without cosmological modeling, for it is not likely that the cosmological ages at two widely displaced and randomly oriented events (relative to the observer) is observed are the same. The proper distance between two cosmologically placed points increases with time as the universe expands. A *comoving* distance is a proper distance divided by the scale factor of the universe. Thus, as the universe expands, the comoving distance remains the same as long as the objects are at rest within the Hubble flow. Distance measures that are somewhat more practical are the luminosity distance, d_{L}, and the angular diameter distance, d_{A}, both of which give corrections to cosmological effects on an object's bolometric luminosity and observed angular diameters, respectively, as a function of the present scale factor and the proper distance. We discuss these measures of the Universe in §3.1.

2.1 Galaxy Interactions

Nearby galaxies now have physical scales (to a moderately low surface brightness level – say, 25 mag/□″) near 8 kpc for a common spiral (Freeman, 1970). This galaxy radius, compared to their mean linear separations of $\overline{r} = (\Phi^*)^{-1/3} \sim 5$ Mpc forms the "impact" ratio of $\simeq 1.6 \times 10^{-3}$. However, in a group environment, an impact ratio of $\sim 4 \times 10^{-2}$ is common.

Of course, this ratio is far higher than the one would construct for typical stars in our portion of the Milky Way; the stellar main-sequence radius is $\sim 10^{11}$ cm compared to a typical separation of $\sim 3 \times 10^{18}$ cm (1 pc), so the impact ratio here is $\sim 3 \times 10^{-8}$. Thus the well-known rarity of stellar collisions of main sequence stars against their "brothers", are rare*..

Still, we must consider changes from a past where the Universe was more dense, and the ratio of sizes of even small "pre-galactic clumps" to their separations were larger and competitive – thus giving gravity a chance to produce a merger product. And there is evidence that such mergers change galaxy morphologies (Genzel et al., 2001; Hibbard et al., 2000; Mihos, 1995; Barnes & Hernquist, 1992). One difficulty which makes the present standard (merger) picture more qualitative than we'd desire, is the fact that the gravitational dimension of a young galaxy depends upon the size and mass-distribution of its omnipresent *dark matter*. These halos may occupy a large physical space – a scale might be a substantial portion of the separations of the galaxies. In a hierarchy of dark matter halos with interior baryonic galaxies, the frequency of major and minor mergers could be a function of galaxy halo mass. That would be a simple evolutionary expectation; it might show up as a series of morphological changes. For some time the Toomres (Toomre & Toomre, 1972) have argued that close encounters of normal galaxies (largely spirals) can generate "mergers" that transform, at least temporarily, the morphological appearance of the baryonic component (see Hibbard et al., 2000). Tidal tails are a likely consequence of a galaxy pair interaction that is relatively close and "slow"; a match between the angular speed of one galaxy and the other's disk (rotational) velocity [$\sim$200 km s^{-1}] will induce effective tides and resonances.

Modern computer codes (see Barnes, 1988; Helmi et al., 1999) which usually include dark matter halos and the self-gravity of the internal baryonic mass points have well reproduced the appearance of the tails for certain well-known merger suspects.

---

*Except perhaps close to the centers of dense, nuclear stellar systems.

FIGURE 1: NGC 4676A,B, taken with the HST Advanced Camera for Surveys (ACS) showing the results of a recent interaction of two spiral galaxies. These galaxies are thought to be on the edge of the Coma cluster. Photo from NASA ACS Science and Engineering Team, NASA and the Space Telescope Science Institute (STScI). (Reproduced in colour in the colour section.)

FIGURE 2: The polar-ring galaxy NGC 4650A; a Hubble Heritage image. These systems are thought to be formed by galaxy interactions; perhaps a "direct hit", in the above figure. The relatively large frequency of polar-ring galaxies suggests that they are dynamically relatively stable, compared to their non-polar cousins. Image courtesy of STScI. (Reproduced in colour in the colour section.)

Interactions and mergers also may well play a large role in the evolution of normal E galaxies, in and away from rich clusters. Barnes & Hernquist (1992) describe several types of galaxy interactions. Their bridges and tails are common in interactions between disk galaxies – they are likely to extract narrow structures of stars and gas. This description is well-represented by Fig. 1, the so-called "mice", NGC 4676A,B. The mergers or "fly-bys" involving E systems are often diffuse-looking; other acquisitions of small companion galaxies (minor-mergers) may lead to shell structures surrounding the surviving dominant E. Other forms of mergers result in a nearly polar ring of gas-rich debris surrounding an early type galaxy, as shown in Fig. 2 (Bournaud & Combes, 2003). Polar rings have long-term stability in the halo of the parent galaxy, and may represent ancient mergers.

Another interesting detail which may be explained by the accretion of a small galaxy with angular momentum are the E galaxies containing kinematically distinct cores, with a central axis quite separate from the rotational direction of the larger E-system as a whole.

In the most extreme examples (see Schweizer & Seitzer, 1992), one can advocate interactions of disk galaxies that eventually yield fairly smooth and regular systems that may even simulate successfully the internal velocity fields of E galaxies. Modeling of a main merger event with two gas-rich disks such as that by Hibbard & Mihos (1995), can satisfactorily reproduce the tails and body of the recent merger NGC 7252 (some 8×10^8 years ago). It might be about half-way towards smoother E surface brightness gradients. If this case is "typical" we can imagine future imagining of definite morphological trends with cosmic epoch,

In the Milky Way galaxy, there is evidence from stars in merger victim streams. For example, red giants with spectra from Sloan Digital Sky Survey were plotted by their radial heliocentric velocity, showing a coherent flow of stars clearly distinct from the Galaxy's numerous disk stars. These atypically orbiting stars were found to be part of a tidally disrupted dwarf galaxy, known as the Sagittarius dwarf, in its last stages of accretion to the Milky Way (see Ibata et al., 2003; Newberg et al., 2002; Mayer et al., 2002; Schweizer, 2000; Helmi & White, 1999; Ibata et al., 1997; Ibata & Lewis, 1994). A similar structure on this side of the galaxy is the mainly gaseous Magellanic Stream, understood in terms of the tidal disruption of the Large Magellanic Cloud (Moore & Davis, 1994).

The search for clues as to the extent and character of galaxy mergers can take several forms. The direct study of massive dark halos at large to small redshifts (corresponding to early and later cosmic epochs) is still impracticable. But we can, in principle, learn about the history of the baryonic component of a composite galaxy by examining the fossil records [motions, abundances] from old stars in our Milky Way, as noted above.

Of course, a history of stellar system mergers might be laid out for us to observe directly with the morphological properties of high-z galaxies; we discuss what few observations are actually available in Chapters 3 and 4. For example there has been discussion and now some observations of the redshift-frequency of bars in the stellar structures of spiral galaxies.

2.2 Evolution of the Stellar Content of Normal Galaxies

Another kind of "systematic" trend we'd expect over the course of cosmic time is the evolution of the stellar content of galaxies. The nuclear evolution of stars is largely a "one-way street", as hydrogen in the stellar core is exhausted. So here a prediction of change with cosmic epoch is quite "safe".

Analysis of the star-formation rate (SFR) in the Milky Way (MW) as a function of its age (Rocha-Pinto et al., 2000) shows an intermittency suggestive of episodic accretion of large quantities of gas, and may help to explain the large dispersion in the metallicity of even recently formed stars at a given galactocentric distance (Edvardsson et al., 1993). Given a galactic stellar mass of 2×10^{10} $M_{\odot}$, the average SFR is ~ 1.5 $M_{\odot}$ yr^{-1}. Rocha-Pinto et al. imply the SFR may have risen from ~ 0.25 $M_{\odot}yr^{-1}$ to an average of ~ 2.4 $M_{\odot}yr^{-1}$ over the last few 100 Myr, though with a large uncertainty.

The traditional view of the formation of elliptical galaxies pictures a monolithic collapse, but as noted by Stanford et al. (1998), this does not sit well with the current models of hierarchical merging scenarios. These models predict a relatively late formation ($z \lesssim 2$) by the merging of smaller galaxies including spirals (e.g. White & Frenk, 1991; Kauffmann et al., 1993). A wide range of ages is possible under this scenario, but color differences "damp out" in time owing to the relatively rapid evolution of the brighter young stars out of the main sequence.

To check for signs of dispersion in ages of the stellar component one could use integrated color magnitude diagrams of the galaxies in the Coma cluster as a standard to compare with high-redshift clusters. If there are systematic age differences in clusters, then the slope and zero point of the color–magnitude (CM) diagram will change with time, only approaching the slope of the Coma CM diagram at low redshift. If clusters are young when seen at high redshift, and there is a systematic gradient in age with galaxy mass, then a passively evolved low-z form would betray itself by the modification of the CM diagram slope. However, recent studies (Blakeslee et al., 2003; Stanford et al., 1998) show that the slope in the E-galaxy color–magnitude relationships are unchanged as one considers higher redshifts (up to $z = 0.9$). This suggests that it is the galaxy mass–metallicity relation that determines the color–magnitude slope, not age differences, whose effects have been damped out with time. This supports the dominance of a passively evolving population with modestly small luminosity evolution ($\Delta_{\rm m} = 1^{\rm m}.5$) and only slight color change ($\Delta_{\rm color} = 0^{\rm m}.1$). Stanford et al. (1998) conclude that galaxies in rich clusters formed most of their stars at high redshift in a well-synchronized fashion, and evolved quiescently thereafter.

An alternative view on the evolution of cluster colors, based on HST/WFPC2 observations of MS2053-04, a massive cluster at $z \sim 0.59$, is presented by Tran et al. (2005). MS2053 is seen experiencing the merger of a somewhat smaller cluster. They find that ~44% of galaxies in the smaller cluster are blue star forming galaxies, while the main cluster has only about 25% blue galaxies. Their analysis suggests that the Butcher–Oemler effect (the observation of the progressively bluer rest-frame colors of galaxy clusters as one probes higher redshifts; Butcher &

Oemler 1978,) is linked to galaxy infall, rather than the aging of a stable stellar population. That is, blue cluster members, and precursors of S0s, are thought to originate in the field. Hierarchical clustering of clusters, it would seem, is on going at $z \simeq 0.6$, and the later cluster members come from a variety of sources.

The E galaxies have had a SF history mainly masked by the "fog of time". Most of the galaxy cluster and field E systems are quite red in color, and their spectra also indicate a light dominance by an old population of stars (Spinrad et al., 1997). Qualitatively this is true over all redshifts available for spectroscopic study with current technologies.

One might envision measuring, or at least modeling, a population mix that could solve for an equivalent stellar age or, more-specifically, the last epoch of major star-formation. The extant evolutionary models (Bruzual A., 2001; Bruzual A. & Charlot, 1993) are well-constrained by good UV and optical region rest-frame spectra which are sensitive to stellar age (and, to a lesser degree, the metal abundance level). This follows from the fact that in the integrated light of a model E galaxy, the stellar main-sequence turn off should play a considerable role in the composite near-UV spectrum, thus giving us an opportunity to age-date a massive single burst of SF in an ideal case.

A reasonable selection criterion that can be used by observers to initially locate moderately distant E galaxies whose evolution might be predictably slow would be galaxy symmetry, plus the very red color of a stellar system with few young and massive stellar components lighting up the emitted UV. Indeed, the spectral region from rest-frame 2000–3000 Å is quite sensitive to the presence of small numbers of late A or early F stars (masses $\sim 1.5\ M_{\odot}$). A main-sequence turnoff in this region implies that the last stage of SF occurred some 2 to 3 Gyr earlier. We pursue this line of reasoning, and consider some consequences in §3.6.

2.3 Evolution of the Gas Mass Fraction

The evolution in the gas mass fraction with time must be intimately connected with the evolution of the stellar content. Since gas is "pre-stellar stuff", we may perhaps naively expect that the gas mass fraction during the initial stages of galaxy-formation should be a systematically greater percentage of the baryons. Clearly this is true at the moment of their formation. However, there is likely to have been a "noisy" decline of the gas mass fraction with time, especially given the evidence for hierarchical merging of gas-rich systems, noted in the above sections. Recall that the aging of early generations of massive stars results in the ejection of some of their nucleosynthetic products into the interstellar medium (ISM), thus affecting the nature of subsequent generations of stars.

Gas-rich galaxies can often be found by their copious star-formation rates. There exist, however, galaxies whose gas mass fraction is high, and have existed (quietly) for a long time, but have produced few stars. That situation may be caused by a large angular momentum. This subject is investigated with the help of the dimensionless spin parameter, λ, proposed by Peebles (1969).

$$\lambda = \frac{J\sqrt{-E}}{GM^{5/2}}, \tag{1}$$

where J is the total angular momentum, E is the total energy, M is the total mass and G is the gravitational constant. While λ is difficult to compute for actual galaxies, it is straightforward for simulated galaxies. Early simulation work (e.g., Burkert & Hensler, 1988) showed that low-λ galaxies tend to be ellipticals, and high-λ galaxies are spirals. For more extreme values of the spin parameter ($\lambda > 0.07$), low surface brightness (LSB) galaxies are predicted (Boissier et al., 2003).

Another way to address the question of the gas in galaxies is by the calculation of the gas depletion timescale – the gas mass divided by the star formation rate. Actively star-forming galaxies generally have timescales of about 2–4 Gyr (Dopita, 1985; van den Bergh, 1991). This in itself is suggestive of substantial replenishment by low metallicity or primordial gas. For dwarf galaxies, gas depletion timescales can be very short, leading to systems which have short episodic bursts of star formation, often with billions of years in between. But, as we have seen, some dwarfs have very long gas-depletion timescales. The story here appears complex.

Models of the chemical enrichment of the Galaxy call for at least partial replenishment of gas supplies by accretion of primordial gas. The episodicity in the SFR (Rocha-Pinto et al., 2000) may indicate that, at least for a "field galaxy" like the MW, the gas mass fraction might be variable, though perhaps with a trend toward lower gas mass fractions with time. This scenario is found consistent with a hierarchical regime of galaxy formation and evolution.

For which phase (ionized, neutral, molecular) of ISM gas do we anticipate the most robust signature of evolution? That question might not have an obvious answer; or perhaps it contains all similar answers! Our next chapter, in fact, emphasizes the apparent evolution of the ionized component of the ISM (to intermediate redshifts). That is because it's easier to observe at modest look-back times; there are practical observational limits to assessing the neutral gas content at $z > 0.2$. We also wish to evaluate the gas content of common spiral systems far into the future – will any ISM remain to clearly show an ionized component? The answer to this question must refer to the sources of replenishment. The population of high-velocity clouds (HVCs), and, beyond them, the low-redshift Lyα clouds, must be prime candidates (Manning, 2003).

2.4 The Chemical Evolution of Galaxies

Another aspect of anticipated change in galaxies are the abundances of heavy elements in their stellar and ISM content. This process begins with the gravitational coalescence of gas. Two galaxy-formation models dominated the landscape of early chemical evolution models. The monolithic collapse scenario (Eggen et al., 1962) produces a burst, then a declining rate of SF, based increasingly on gas supplied by evolved stars. The result is a rather rapid increase in metallicity. Such a model is called a closed-box approximation (Tinsley & Larson, 1978). It was initially and wa applied to both ellipticals and spirals (with the addition of angular momentum in the protogalactic cloud Larson 1976). A problem with such monolithic models is that many small stars of very low metallicity ought to have been formed, and should still be around today. This is known as the G-dwarf problem – the problem being the apparent lack of a large population of extremely low metallicity sub-dwarfs. An alternate model of early evolution was postulated by Searle & Zinn (1978), which may be classed as an early hierarchical model of galaxy formation. This model produces a more slowly increasing galactic metallicity. In such a model, only a few "G-dwarfs" will be formed before the nucleosynthetic products of large stars are spread throughout the galaxy.

We find very few old stars with $[M/H]$ ratios below 10^{-4} of the solar level in our Galaxy. Do we anticipate some young and distant galaxies with overall metallicities down to an extremely low level like that value? Such chemically young galaxies would provide a potentially useful laboratory in which to understand the chemical, stellar, and structural evolution of galaxies (Phillips et al., 1997).

In the very general cosmological context of 3-D simulations, the evolution of metallicity as a function of density and time may be modeled. Simulations show that dense regions evolve their metals more rapidly than low-density regions, quickly reaching an abundance saturation point at approximately a solar value (Cen & Ostriker, 1999). Low-density regions have few metals because stars form inefficiently there, and because the transport of metals to isolated regions is inefficient.

Since supernovae (SNe) of various types are expected to differentially pollute their environs with metals of the oxygen group and the iron-peak in the periodic table of elements, we'd even imagine some evolution in (O/Fe) abundances as functions of cosmic epoch (Freeman & Bland-Hawthorn, 2002; Hensler, 1999). SNe of Type II (from massive stars) are expected to initially be enriched in the "α" elements: those with even atomic weights up to 48, such as O, Mg, Si, S, Ca, and Ti. These SNe should be followed by a gap of $\sim$1 Gyr before the lower-mass SNe Ia begin to play a dominant role in enhancing the iron-peak elements. Thus, one would expect, as indeed we have discovered locally, that the enhancement of the α elements is progressively stronger as lower metallicity environments are probed, such as the stars in the Milky Way halo (McWilliam, 1997).

When looking at the evolution of the metallicity of the galactic structural components (disk, halo and bulge), we need to recognize that enriched gas is expected to be copiously expelled from them during the epoch of their formation. Thus, they enrich not only the gas from which they are being made, but also perhaps the other spatial components. For instance, enriched halo gas may pollute the bulge stars, and the later-forming disk. Accurate modeling will require a self-consistent chemodynamical analysis to correctly model the evolution of these stellar populations (Hensler, 1999). Such modeling also requires, for concordance, the accretion of quantities of primordial gas.

Again, in Chapter 3, we'll consider a variety of attempts to measure abundances in classic galaxies and other "objects" at moderate to high redshift, and compare the parameters to local values. Here we are somewhat handicapped by the fact that we don't know which high-z objects are progenitors of local galaxies, and by what process the transformation comes about. Moreover, the redshift itself imposes practical boundaries to certain observations which are very useful at observed wavelengths and low redshifts utilizing CCD detectors. However, visual emission lines redshifted to the IR for $z > 1$ become less useful as they then are observable only with technologically less-mature detectors (e.g., bolometers), ratjer than CCDs. CCDs are clearly the detectors of choice below 1 μm, and much active research is being invested into this challenging subject.

2.5 The Chemical Evolution of AGN at Moderate Distance

Our expectations concerning the evidence for evolution of active galactic nuclei (AGN) are less mature than other evolutionary effects we anticipate. That's not too surprising, as our physical understanding of massive black holes (MBH) in nuclei is very recent. Since our expectation involves the concept of the MBH supplying gravitational potential energy, evolution could take place in the MBH mass, the MBH near-environment, and thus the accretion rate, or even in its plurality. But there does not exist a well-founded dominant expectation accepted by the field; lots of questions are brewing.

AGN are now understood to be active phases of galaxy nuclei; an MBH with the accretion disk fulfilling the intermediary physical position between the MBH and the larger galaxy, at least on a logarithmic scale. Thus, galactic nuclei come in two classes; active and inactive, or quiescent. The comoving density of MBHs must increase monotonically with time, but the fraction which are active appears to decline.

An observational correlation which is driving most attempts to understand MBH activity (i.e., AGN) is what is known as the Magorrian relation, $M_{\rm BH} = 0.006\, M_{\rm bulge}$ Magorrian et al. (1998). For a massive bulge ($M_{\rm bulge} \geq 1.\times 10^{11}$ M$_\odot$), black hole mass can reach a billion M$_\odot$. There also exists a fairly tight relationship between the mass of the black hole and the mass, or velocity dispersion of the bulge. Note the $M_{\rm BH}$-σ relation,

$$M_{\rm BH} = M_\odot \left(\frac{\sigma^*}{200 \text{ km s}^{-1}} \right)^{4.02} , \tag{2}$$

where the power, 4.02, is a finding of Tremaine et al. (2002) (see Fig. 3). There also exist relationships with other indicators, such as the spectral width of the broad-line region, attributable to the velocity dispersion of close-in clouds whose circular motion is dominated by the BH (Shields et al., 2003).

The "activity" of the MBH appears connected with the existence of an accretion disk whose viscous differential rotation heats the gas to high temperatures, producing a spectrum well-fit by a power law. For MBHs, there is a radius at

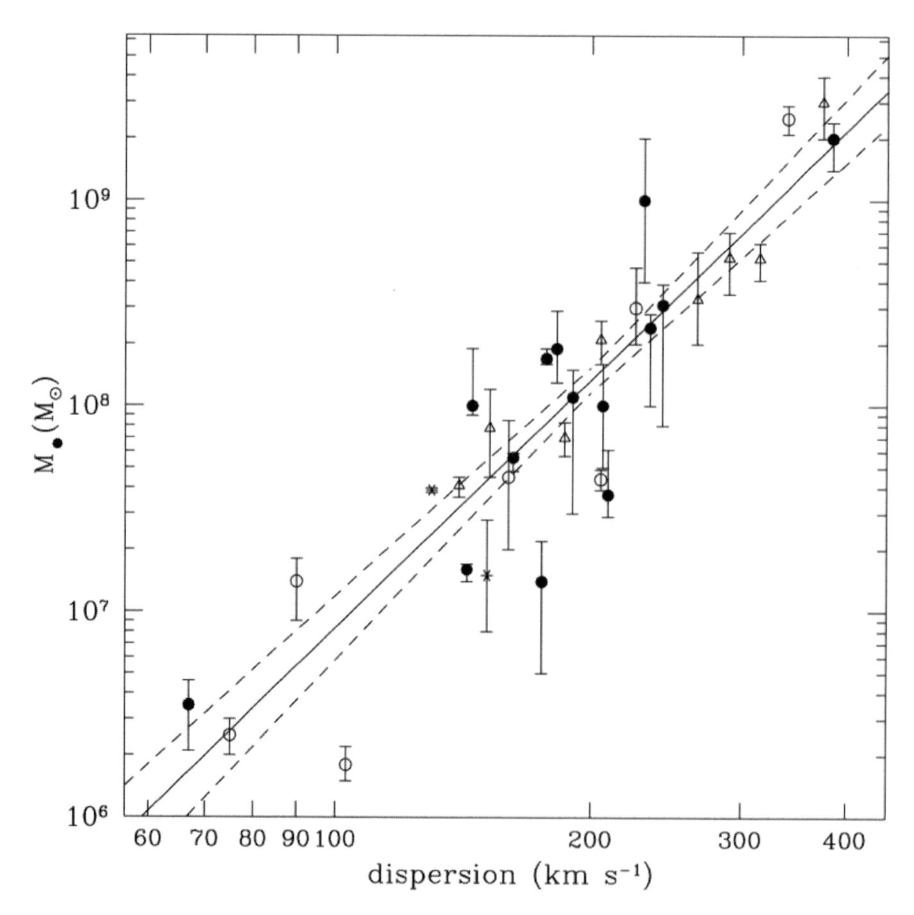

FIGURE 3: Data on black hole masses as a function of stellar velocity dispersions. The fit is the same as Eq. (2). Mass measurements based on stellar kinematics are denoted by circles; gas kinematics are denoted by triangles, and maser kinematics are denoted by asterisks. Filled circles are results of Gebhardt et al. (2000), which use stellar velocity dispersions. The dashed lines show the 1-σ limits on the best-fit correlation. Figure from Tremaine et al. (2002).

which the orbital velocity is that of light. This provides a limit to the inner radius of the accretion disk. The mechanisms by which gas is moved toward the accretion disk may be numerous. A rapid movement of gas toward the nuclear region is often associated with major merger events. This probably involves a number of different mechanisms such as violent relaxation (Lynden-Bell, 1967), dynamical friction (Chandrasekhar, 1943), and hydrodynamical shocks with efficient cooling. Violent relaxation is a rapid smoothing of the matter distribution when the gravitational potential changes on timescales comparable to the dynamical timescale. Dynamical friction is the deceleration of clumps of matter falling into the smoothed matter density of a larger system. The convergent movement of matter in the larger system toward the interloper causes an enhanced density field in the lee of its path, causing a deceleration: dynamical friction.

These large-scale events involve an exchange of momentum within the galactic components. The exchange of angular momentum has the tendency to funnel mass inward and carry angular momentum outward; a function also served by the more orderly density waves in spiral galaxies.

Mergers may also stimulate the activity of the AGN by providing a new supply of gas to the nuclear zone. The adjustments of pressures and flows would plausibly cause an enhanced flow into the accretion disk (Kewley & Dopita, 2003; Hatzimi-naoglou et al., 2003). One approach to modeling the probability of an AGN being active as a function of epoch considers the gas mass fraction in the bulge during its formation (Di Matteo et al., 2003). Another method is to note the rapid growth of metallicity in quasars, a semi-independent sign of violent stellar evolution, which may be associated with the causes of AGN activity (Wang, 2001). Major mergers may be considered as likely causes themselves, for if the Magorrian relation ($M_{\rm BH}$-σ) is fixed by the bulge mass, a major merger will change the bulge mass, "requiring" accretion to the MBH in order to maintain the M–σ relation. The real reason is not likely to be so ad hoc. The violent relaxation involved in such mergers will cause a randomization of stellar orbits. Some of these will be plunging orbits that will put them too close to our MBH. These stars may be disrupted and swallowed by the MBH, after which the activity may decline. Nuclear activity is probably often a combination of all of these mechanisms.

The result of a QSOs activity may be it's own demise, as the energy released may sweep up interstellar gas and actually eject it from small to moderate-sized galaxies (Silk & Rees, 1998).

These ideas may lead to predictions for the evolution of QSOs, which must seek observational confirmation. Were some QSOs more luminous in the past (i.e., $z > 1$) than recently? This could be physically expressed as the comoving luminosity density. If the QSOs luminosity density evolves in comoving coordinates, why, and how does it evolve? Can the "host galaxies" help explain the nature of any observed QSO time-evolution? Is the galaxy merger rate a dominant influence on the near-nuclear environments? Our physical models seek coherence with observation within the standard model of cosmology. One example can be found in Di Matteo et al. (2003), who use prescriptions for black hole growth and host galaxy metallicities to model quasar environs and the growth of the BH (and the decline in SF with cosmic epoch).

2.6 Summary

In this chapter we wondered what properties of the Universe might conceivably change in time, as a means of anticipating and preparing the way for the findings presented in Chapter 3.

Because the Universe appears to be going from a denser to a less-dense state, it follows that the merger rate of galaxies, and possible sub-galactic structures, may have been higher in the past than in the present, especially if the Universe has a cosmological constant, an accelerating component to the expansion of the

Universe. This parameter is introduced in Chapter 3, and discussed in some detail in Chapter 5.

The stellar content certainly evolves in ways specific to galaxy type. For E-galaxies, this is especially true as they are usually well-modeled as being formed by a single burst of star-formation. In these systems, massive stars progressively evolve off the main sequence, leaving cooler stars to dominate the spectral energy distribution (SED). For spirals and irregulars, the star-formation rate may trail off, hold steady, or even increase with time, resulting in SEDs that are bluer, with lower mass–luminosity ratios.

The gas mass fraction may also change, according to the SFR, and the available gas. In a "closed box", neutral gas would disappear fairly rapidly in spirals. However, mergers may bring in new quantities of gas, thus moderating its depletion.

The chemical evolution (abundances of metals) will be closely linked to the evolution of the gas mass fraction, for the latter represents the possibility for the formation of the rapidly evolving O and B stars, and hence the enrichment of the gas. Again, this evolution may be moderated by accretion of primordial gas.

We may also expect the activity of MBHs to vary with time. If AGN activity is related to mergers, or to the SFR near the nucleus, then we may expect AGN activity to vary with the merger rate and the star formation rate.

REFERENCES

Barnes, J. E. 1988, ApJ, 331, 699

Barnes, J. E. & Hernquist, L. 1992, ARAA, 30

Blakeslee, J. P. & 22 co-authors. 2003, ApJL 596, L143

Boissier, S., et al., 2003, MNRAS, 343, 653

Bournaud, F. & Combes, F. 2003, A&Ap, 401, 817

Bruzual A., G. & Charlot, S. 1993, ApJ, 405, 538

Bruzual A., G. 2001, Astrophysics and Space Science Supplement, 277, 221

Burkert, A. & Hensler, G. 1988, A&Ap, 199, 131

Butcher, H. & Oemler, A. 1978, ApJ, 219, 18

Cen, R. & Ostriker, J. P. 1999, ApJL, 519, L109

Chandrasekhar, S. 1943, ApJ, 97, 255

Di Matteo, T., et al., 2003, ApJ, 593, 56

Dopita, M. A. 1985, ApJL, 295, L5

Edvardsson, B., et al., 1993, A&Ap, 275, 101

Eggen, O. J., et al., 1962, ApJ 136, 748

Freeman, K. & Bland-Hawthorn, J. 2002, ARAA, 40, 487

Freeman, K. C. 1970, ApJ, 160, 811

Gebhardt, K., et al., 2000, ApJL, 539, L13

Genzel, R., et al., 2001, ApJ, 563, 527

Hatziminaoglou, E., et al., 2003, MNRAS, 343, 692

Helmi, A. & White, S. D. M. 1999, MNRAS, 307, 495

Helmi, A., et al., 1999, Nature, 402, 53

Hensler, G. 1999, Ap&SS, 265, 397

Hibbard, J. E. & Mihos, J. C. 1995, AJ, 110, 140

Hibbard, J. E., et al., 2000, AJ, 119, 1130

Ibata, R. A., et al., 2003, MNRAS, 340, L21

Ibata, R. A. & Lewis, G. F. 1994, Nature, 370, 194

Ibata, R. A., et al., 1997, AJ, 113, 634

Kauffmann, G., et al., 1993, MNRAS, 264, 201

Kewley, L. J. & Dopita, M. A. 2003, in Active Galactic Nuclei: from Central Engine to Host Galaxy, meeting held in Meudon, France, July 23-27, 2002, Eds.: S. Collin, F. Combes and I. Shlosman. ASP (Astronomical Society of the Pacific), Conference Series, Vol. 290, 519

Larson, R. B. 1976, MNRAS, 176, 31

Lynden-Bell, D. 1967, MNRAS, 136, 101

Magorrian, J., et al., 1998, AJ, 115, 2285

Manning, C. V. 2003, ApJ, 595, 19

Mayer, L., et al., 2002, MNRAS, 336, 119

McWilliam, A. 1997, ARAA, 35, 503

Mihos, J. C. 1995, ApJL, 438, L75

Moore, B. & Davis, M. 1994, MNRAS, 270, 209

Newberg, H. J., et al., 2002, ApJ, 569, 245

Peebles, P. J. E. 1969, ApJ, 155, 393

Phillips, A. C., et al., 1997, ApJ, 489, 543

Rocha-Pinto, H. J., et al., 2000, ApJL, 531, L115

Schweizer, F. 2000, Royal Society of London Philosophical Transactions Series A, 358, 2063

Schweizer, F. & Seitzer, P. 1992, AJ, 104, 1039

Searle, L. & Zinn, R. 1978, ApJ, 225, 357

Shields, G. A., et al., 2003, ApJ, 583, 124

Silk, J. & Rees, M. J. 1998, Ap, 331, L1

Spinrad, H., et al., 1997, ApJ, 484, 581

Stanford, S. A., et al., 1998, ApJ, 492, 461

Tinsley, B. M. & Larson, R. B. 1978, ApJ, 221, 554

Toomre, A. & Toomre, J. 1972, ApJ, 178, 623
Tran, K. H., et al., 2005, ApJ, 619, 134
Tremaine, S., et al., 2002, ApJ, 574, 740
van den Bergh, S. 1991, PASP, 103, 609
Wang, J. 2001, A&Ap, 376, L39
White, S. D. M. & Frenk, C. S. 1991, ApJ, 379, 52

Chapter 3

Observations of an Evolving Universe

Here we examine some departures from the anticipations of a stable and unevolving Universe of galaxies. We need not penetrate to the most distant "Island Universes" to notice these differences; indeed, here we'll mention some effects at moderate redshift which may include those anticipated in Chapter 2, but may also be unanticipated and only partly physically understood. Many of our recent advances are due to the upgraded acuity of the HST.

3.1 The Metric Measures of an Evolving Universe

As mentioned in the previous chapter, the distance measures which are most helpful in understanding physical quantities extracted from high-redshift data are the luminosity distance, d_{L} and the angular diameter distance d_{A}. To understand the evolution of the past Universe, we need to know what these measures mean in the present day units of length and time. To do this we must bear in mind that the answers will vary according to which cosmological model we choose to adopt.

The luminosity distance d_{L} is the distance an object would have to be at for the bolometric flux to be equal to the luminosity divided by $4\pi d_{\mathrm{L}}^2$ in a flat, static universe. The angular diameter distance d_{A} is the distance an object has to be placed for its angular displacement in the sky to be explicable in terms of its true size divided by d_{A} in a flat, static universe. The relation $d_{\mathrm{L}} = d_{\mathrm{A}} * ((1+z)^2$ holds for all cosmologies (Weinberg, 1972). The luminosity distance becomes much greater than the angular diameter distance at high redshift. While the luminosity distance always increases with redshift, the angular diameter distance sometimes reaches a maximum, then begins to decline again – as, for instance, in the critical density Einstein–de Sitter model (see Chapter 5). This is due, very generally, to the lensing effect of the Universe as a whole.

To understand the concept of comoving, in relation to proper volumes, we present the following example. Assume there are eight galaxy clusters seen at some redshift z, and placed at rest at the vertices of a cube. The enclosed volume can be expressed as a proper volume, or as a comoving volume. The *proper* volume is then describable as the cube of the proper distance along the cube edge, l^3. The proper distance increases with the scale factor, and the scale factor varies as $(1+z)^{-1}$, so the proper volume varies as $(1+z)^{-3}$. Thus, the proper number density of the clusters decreases with time, $n = ((1+z)/l)^3$ (there is only one-eighth of each cluster in our cube; the other seven-eighths are in the other seven cubes that intersect a given vertex in our idealized universe). The *comoving* volume of the structure seen at redshift z is defined as the volume it would have if it was seen at the current cosmological epoch (defined by $z = 0$). Seen at redshift z, the proper volume is smaller than the comoving volume by a factor of $(1+z)^{-3}$. The comoving number density of a non-evolving population, such as our galaxy clusters above, would be constant in time, $n = l^{-3}$. Comoving volumes enable easy comparison of measures of distant volumes and of large-scale structure to that of local structure.

Figure 1 shows the look-back time in gigayears as a function of redshift, where relations for three standard cosmological models are shown. The dotted line represents the standard Einstein–de Sitter model (a standard, "critical density" Big Bang), the dashed line is an open (empty) Big Bang, and the solid line represents the current standard model, a spatially flat cosmological constant model, which includes a component of exponential expansion. The horizontal dotted, dashed and solid lines show the limits of look back time as redshift approaches infinity for the three models. The Einstein–de Sitter model has a universe age less than 10 Gyr; in conflict, we shall see, with globular star cluster ages. We discuss the currently accepted flat-Lambda model in much more detail in Chapter 5. It is generally a good bet that at $z = 1$ the look-back time is about 7 or 8 Gyr when $\mathrm{h} = 0.7$.

Figure 2 shows the differential comoving cosmological volume in the Universe per square arcminute per unit redshift as a function of redshift for three standard cosmological models. Here, the differences between the different models is much more dramatic than in Fig. 1. We have abbreviated the models: a flat $\Lambda = 0.7$ model assumes $\Omega_\mathrm{m} = 0.3$, as in Fig. 1, the *open* model has $\Omega_\mathrm{m} = 0.3$ and $\Lambda = 0$, and the Einstein–de Sitter model has $\Omega_\mathrm{m} = 1.0$ and $\Omega_\Lambda = 0$. The Hubble constant is $70\ \mathrm{km\,s^{-1}\,Mpc^{-1}}$ in all cases. These models allow interpretations of high-redshift data. A standard-sized object placed at high redshift would appear much smaller in a flat Λ universe than in an Einstein–de Sitter universe as it would be immersed in a much larger universe.

3.2 What is New in Galaxy Counts?

Ellis (1997) shows that the simple counting of galaxies as a function of apparent magnitude (B) can reveal useful information about galaxies. Comparing this work with non-evolving models of normal galaxies with a local (Schechter) luminosity

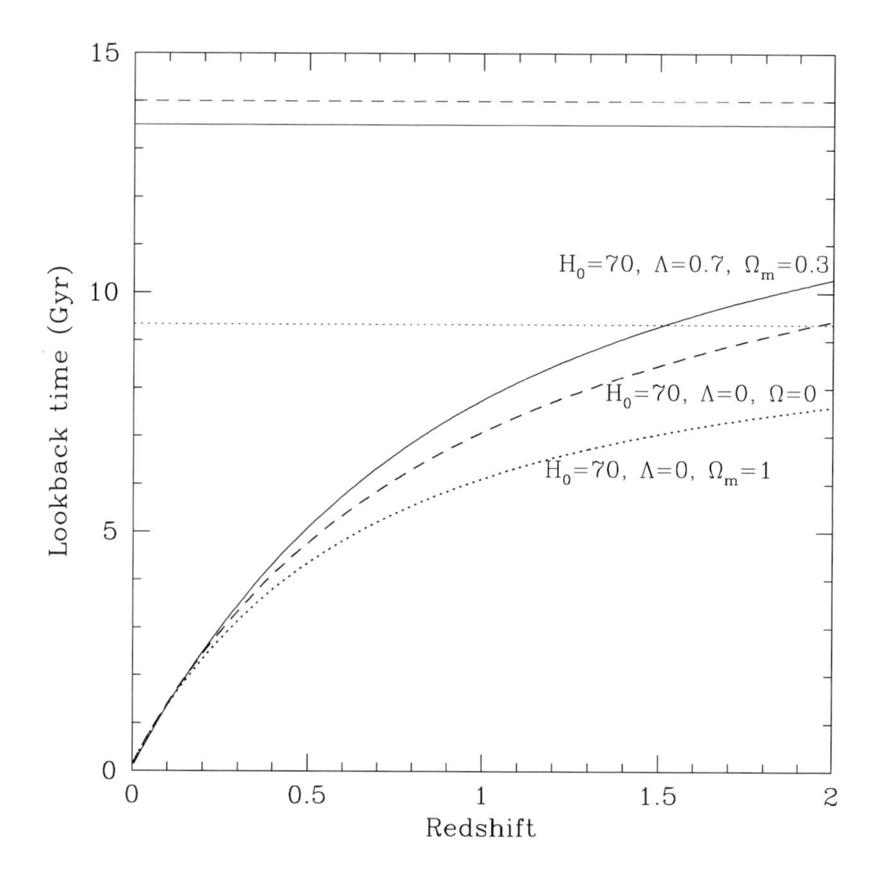

FIGURE 1: Look-back times as a function of cosmology (see annotations) and redshift. For infinite redshift, the look-back times become ages of the universe. These age limits are shown as the horizontal lines whose line types correspond to those of the labeled curves. For the popular Λ model (LCDM; $\Omega_{\rm m} = 0.3$, $\Omega_\Lambda = 0.7$), the age is 13.5 Gyr if $H_0 = 70\ {\rm km\,s^{-1}\,Mpc^{-1}}$.

function and normalization beyond $K = 18$ or especially $B = 20$, there is a noticeable and increasing excess of blue galaxies. This was known and partially accepted by a puzzled community as due to some sort of "evolution", which was well defined in the review of Koo & Kron (1992) (see their Figure 1). They provide a good comparison of predicted no-evolution models and observations of surface density as a function of the apparent magnitude in the K, r, and b_j wave-bands. No-evolution models are based on the local luminosity functions (LF) projected into redshift space, with appropriate corrections for cosmology, and shifting bandpasses (K-corrections). The divergence of observations from the

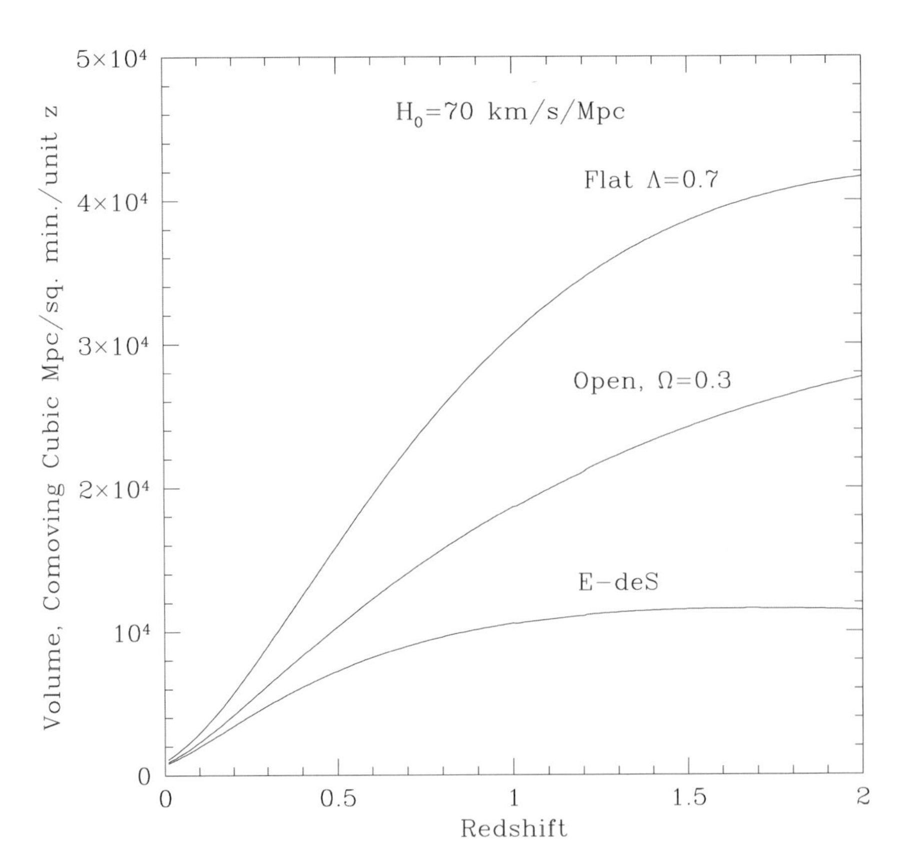

FIGURE 2: Differential comoving volumes as a function of redshift for three cosmologies, the flat Λ model, the open, $\Omega_{\rm m} = 0.3$ model, and the Einstein–de Sitter model.

non-evolving models is known as the "faint blue galaxy problem".

Notice, from Fig. 1, that the slope of the look-back time function is steepest at low redshift. Redshifts $z \lesssim 1$, though apparently small, are thought to comprise a majority of the age of the Universe. It is not surprising, therefore, that there could be dramatic changes in the galaxy population over the last unit redshift.

Utilizing the morphological data available now with HST imaging and the large number of galaxies so surveyed by a long-term parallel survey (see Cohen et al., 2003), the no-evolution theory will fall short of the morphology-split counts for all types; Cohen et al. (2003) point out that all galaxy types show excesses in their counts for $24 \leq B \leq 27$ mag, as can be seen in Fig. 3, which corresponds to a range of *median* redshift $z \approx 0.5$ to 1.5. But they note that the excess over

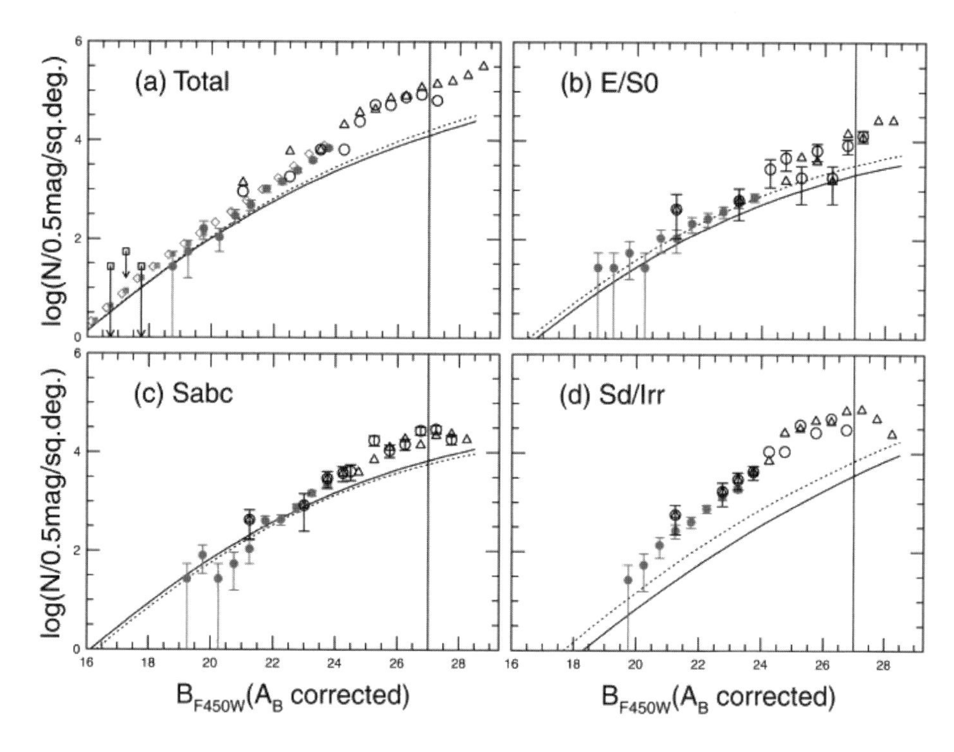

FIGURE 3: Galaxy counts as a function of $A_{\rm B}$ magnitude for (a) all galaxies, (b) E/S0, (c) Sabc, and (d), Sd/Irr. The figure is from Cohen et al. (2003). Two versions of non-evolving models are shown as dotted and solid lines (Marzke et al. 1994 and Marzke et al. 1998, respectively). Evolution is most apparent for Sd/Irr galaxies.

the no-evolution models (lines) grows from a factor of 3 to 4 for early galaxies to a factor of 6–10 for the later morphological types (Sd, Irr). A later graph in Cohen *et al.* (their Figure 4) superimposes evolution models based on results of attributing to galaxies luminosities enhanced by the factor $(1+z)^{\beta}$ $(\beta \geq 1)$. This means small galaxies, which are more numerous, have the luminosity of nearby more massive galaxies, which are less numerous. For E/S0, $\beta = 3$ is somewhat consistent with observations, but for Sd/Irr, even $\beta = 4$ to 6 is not sufficient to explain the observations, as a clear excess remains at $B \gtrsim 24$ mag.

The faint-end slopes of the E/S0, and spiral samples in Cohen *et al.* are near $\alpha = -1$, but the Sd/Irr have $\alpha \simeq -1.8$ (Marzke et al., 1998). What is the source of the problem? Could the volume element used in projecting no-evolution models be incorrect? Are Sd/Irr galaxies evolving into spirals in time, or perhaps fading to obscurity in the local Universe? The slopes of the galaxy count functions in Fig. 3 are steeper than the non-evolving models for all galaxy types, particularly at faint magnitudes $B_{\rm F450W} \gtrsim 24$. One proposal seeks to explain the blue excess by proposing a population of episodic bursting then fading dwarf galaxies at $z \sim 0.5$ to 1.5 (Ferguson & Babul, 1998; Babul & Ferguson, 1996; Babul

& Rees, 1992). Ferguson & Babul conclude that the fading dwarf model predicts many faded dwarfs that should be detectable in the Hubble deep field (HDF) as LSB red objects, but aren't. Even if fading dwarfs could explain the faint blue galaxy excess, this scenario could not explain the excess in grand-design spirals and ellipticals seen in faint magnitudes in Fig. 3 – types which would not be likely to display similar episodic burstings. Notice in Fig. 2 that a flat Λ model has a larger comoving volume per unit z at high-redshift. This is the same model used in the production of non-evolving models in Cohen et al. (2003). If a model with lower comoving volume was used, then the counts at faint magnitudes would be lower. Thus, if we were to fix our problem with the volume element or cosmology, we would need one with a greater volume than a $\Lambda = 0.7$ model at $z \gtrsim 1$. That is unlikely.

How do we presently understand the counts and "count-excess" integrated with the morphological differentiation over a moderate range of redshift? One possible explanation is that faint systems at $z \approx 1$ to 2 may really be smaller and more numerous than similarly luminous galaxies seen at the present epoch, as hierarchical scenarios in general suggest. For consistency, at the faint end of the counts ($z \geq 1$) we should see pre-galactic scale building blocks and merging systems with enhanced SF.

This concept should provide excess counts for all three coarse morphological types used by Cohen et al. (2003) – E/S0; Sa, Sb, Sc spirals; and Sd/Irr.

It may be that the above transitionary scheme creates more "train wreck" Irr galaxies at $z \sim 1$ than it produces bulge-dominated E/S0s, so that the latter smoother morphologic class shows only a modest count excess at faint levels. Or perhaps many bulges formed early and show no further major events that produce luminous stars or any active SF.

However, the observations of van den Bergh et al. (2000) and Conselice et al. (2003) show that massive galaxies have large merger rates (both major and relatively minor mergers) at $z \approx 2.5$, based on an asymmetry criterion (see below) from HST imaging. They find a redshift dependency $R \propto (1+z)^{\mathrm{s}}$ for luminous galaxies, where $s \simeq 4$ to 6. Small galaxies have significantly smaller merger rates. Thus, at high redshifts, the smaller "units" are apparently not growing much by accretion, but they are contributing to the growth of larger galaxies. The merger of small galaxies to larger galaxies has the effect of reducing the slope of the mass function of galaxies with time – reducing the number of small galaxies, while raising the masses of formerly modest-sized galaxies.

For the preferred cosmological parameters of $\Omega_{\mathrm{m}} = 0.3$, $\Omega_\Lambda = 0.7$, the Λ-driven acceleration of the Universe's expansion acts to reduce the expected galaxy merger rate in the present epoch, $z \lesssim 0.5$. The local galaxies, which would thus experience fewer minor mergers in recent times, and would be less prominently luminous, and appear at fainter magnitudes than expected, depopulating the fainter-end of the LF. Looking at it in reverse, the local luminosity function, placed at an elevated redshift ($z \sim 1$) would have fewer sub-L* galaxies than the ambient environment. This scenario is in concert with the observed steady and strong decrease in the comoving SFR density observed for $z \lesssim 2$ (Madau et al., 1998).

Future studies of the counts and merger histories at intermediate and rather low redshifts ($z \leq 0.5$) may elucidate the more subtle effects of Λ on the interaction rates of common galaxies and the change in comoving volume with cosmic epoch (see Cohen et al. 2003).

3.3 The Sizes and Morphologies of Galaxies at Fairly High Redshift

For a quantitative look at the faint field galaxy sizes we turn to Ferguson et al. (2004); the galaxy size is a simple parameter that can be "machine-extracted" for even faint samples. The problem is that the $z \geq 3$ galaxy population appears small in angular size. Even with the new Advanced Camera for Surveys (ACS) imager on the HST the number of pixels involved in the images are modest, for with $0''.046$ per pixel, the half-light radii of $\overline{z} = 3$ galaxies are $\sim 2.0\,\mathrm{h}_{70}$ kpc, as shown in Fig. 4, or $\sim 0''.3$. This leaves a little over 12 pixels across the half-light diameter of the galaxy (Ferguson et al., 2004). The proper (i.e., *not* comoving) galaxy diameter at this redshift is only ~ 4.64 kpc, with $h = \Lambda = 0.7$. To make matters more difficult, it is found that mean galaxy half-light proper diameters decline with increased redshift, going from 6.6 kpc at $z = 2.3$ (the mean survey redshift) to 4.64 kpc and 3.15 kpc at $z = 3$ and 5, respectively. Thus, for the symmetric-in-appearance distant systems, there is only modest morphological data that can be extracted for comparison with galaxies at lower z. We discuss here the morphologies of galaxies at $z \leq 3$, saving our discussion of the increasing fraction of physically small and compact systems seen at higher redshifts for the following section.

As the counts previously suggested, the most striking studies of moderately distant galaxies show that many are smaller than anticipated, and the fewer well-resolved systems convey an impression of a chaotic structure – a larger percentage of interacting, irregular, or peculiar morphologies than we observe nearby. This is true for redshifts greater than 0.5. As Abraham et al. (1996) pointed out, for magnitudes fainter than 25 in I, the conventional and *subjective* Hubble classification scheme provides only a partial description of faint galaxy morphologies. van den Bergh et al. (2000) similarly note a declining fraction of grand-design spirals beyond $z \sim 0.3$.

Locally, it was found that asymmetries in the light distribution of field spirals are likely due to episodes of localized star formation. Local instances of this have been attributed to the accretion of clumps of low-metallicity gas (Zaritsky, 1995). This observation was later formalized in terms of an asymmetry measure (the term used is "lopsidedness") determined by the Fourier decomposition of the disk surface brightness. In practice, the lopsidedness is then the ratio of the $m = 1$ Fourier amplitude to the $m = 0$ Fourier amplitude (the surface brightness) at a radius between 1.5 and 2.5 scale lengths, $\langle \mathrm{A}_1 \rangle$ (Rix & Zaritsky, 1995; Zaritsky & Rix, 1997). They found that about 30% of local spirals have significant asymme-

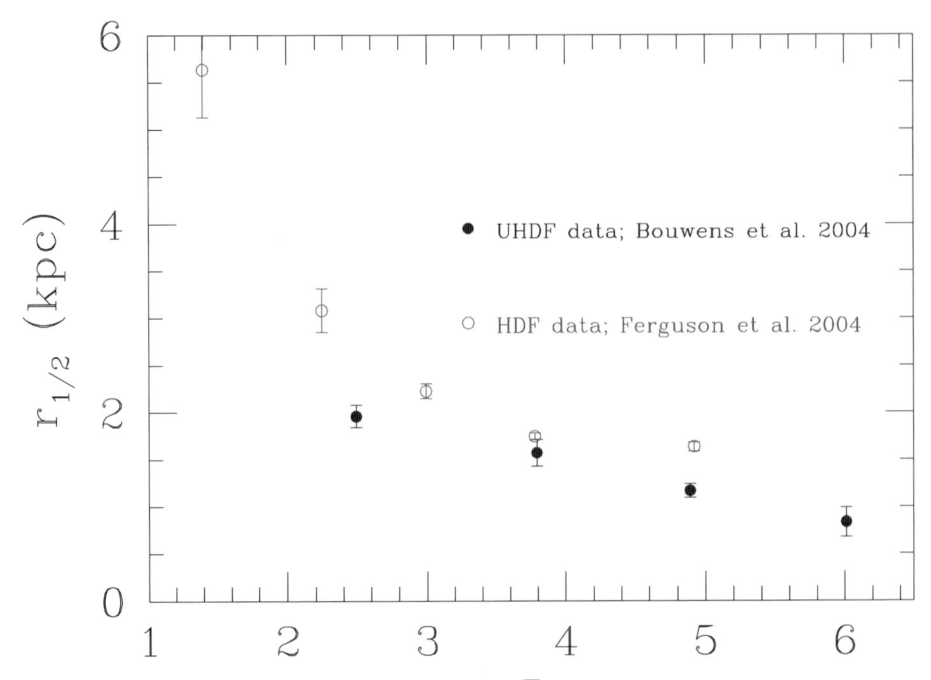

FIGURE 4: Half-light radii as a function of redshift (points with error bars) for a flat Λ model (see text). Open points are from Ferguson et al. (2004), using HDF data, and filled points are from Bouwens et al. (2004), using data from the Hubble's Ultra Deep Field (UDF). In both cases, half-light radii are seen to decline significantly with increased redshift.

tries ($\langle A_1 \rangle \gtrsim 0.2$). In addition, a strong correlation was found (96% confidence) between lopsidedness and the *excess* brightness ΔB, relative to expectations for their circular velocity (i.e., the Tully-Fisher relation), suggesting a relationship between what causes lopsidedness and what affects its luminosity. It is suggested that accretion events – minor mergers – are responsible for the correlation. An estimate of the current probability of an accretion event of a magnitude that would produce significant lopsidedness (estimated at $\sim$ 10% of parent galaxy mass) is between 7% and 25% per Gyr. It would be interesting to see how accretion rates vary with redshift. While this observational technique is not efficient at high redshift (the galaxy image will be under-sampled), we shall see that other substitute measures can be taken.

Shortly following the installation of the HST WFPC2 Camera, several investigation teams noted the peculiar morphologies of faint galaxies, and the "shortage" of classic spirals at redshifts of $z > 1$. Attempts to still utilize conventional classifications on the Hubble scheme have been published; in particular, the work of van den Bergh et al. (2000) illustrates the difficulties in placing some moderately distant ($z \sim 1$) galaxies in the classic ($z \sim 0$) Hubble sequence. The van den Bergh

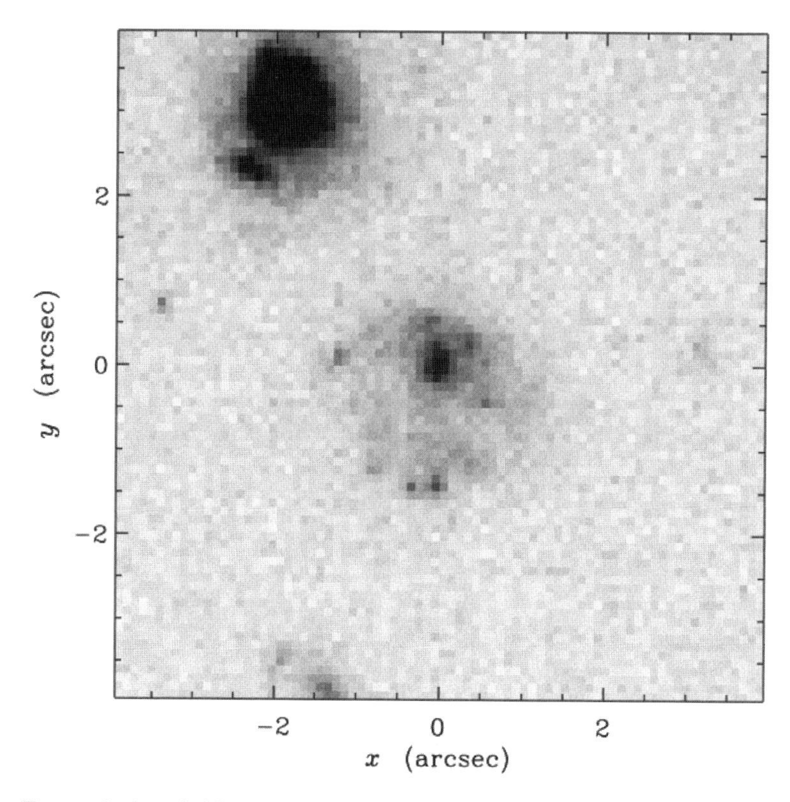

FIGURE 5: A probable protospiral at $z = 0.95$, from van den Bergh et al. (2000). The dense core is composed of old (red) stars, while the outer knots are young and blue. Thus, many distant galaxies avoid simplistic classification.

team classified 241 galaxies in the HDF and Flanking Fields with spectroscopic redshifts covering $0.25 < z < 1.2$. Their classifications utilize previous experience at low z; they are tradition-bound and *subjective*, but consistent, and utilize considerable previous experience in galaxy morphological sorting. Extending the observations of Brinchmann et al. (1998) and Abraham et al. (1999), and Cowie et al. (1995a) with a complete redshift sample, the van den Bergh group classifiers noted,

- Most intermediate and late-type spirals with $z \gtrsim 0.5$ have morphologies less-regular than nearby Hubble types (e.g., see Fig. 5). Spirals with long, and symmetric continuous arms are rare at $z > 0.3$. Even less-pronounced spiral structure becomes less dominant in the distant half of this sample.

- On the other hand a substantial fraction of galaxies in the HDF appear to be involved in interacting and merging systems (e.g., see Fig. 6).

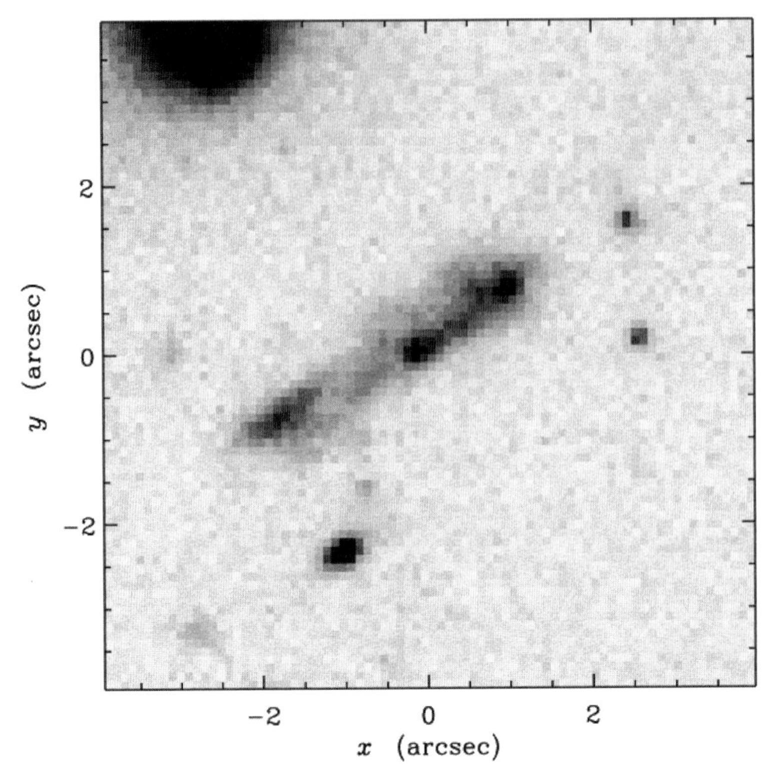

FIGURE 6: A peculiar disk-like object seen edge-on at $z = 1.020$, with three apparent merging sub-galactic components. HST image from van den Bergh et al. (2000)

The above trends have already been corrected, or partly ameliorated by selecting and observing in bands whose rest-wavelengths, at the galaxy redshift, were close to the visible, thereby obviating the need for morphological K-corrections. But generally, the Hubble form diagram, as established from the sample of local galaxies (Fig. 2), becomes increasingly irrelevant as higher redshifts are probed.

It is also possible, and likely advantageous, to make galaxy morphological categorization more impersonal and more numerical by quantitative morphological analysis. However, the resolution of galaxy morphologies are limited by the image size, the angular resolution and the pixel dispersion of the camera which would detect it. Even with long HST integrations (Williams et al., 1996; Giavalisco et al., 2004) in various "Deep Fields" the average galaxy at $z > 5$ shows only marginal resolution. But at somewhat smaller redshifts, especially using the new Advanced Camera for Surveys (ACS) we can analyze some faint and distant galaxy properties that may lead to improved physical understanding of systems in a young Universe. This means "dissection" of faint, small galaxy images.

A program of "physical morphology" (Conselice, 2003) coupled with improved photometric redshifts ("photo-zs") for faint galaxies, has shown much recent progress in our documentation of the evolution of galaxy morphologies from quite high redshifts. This program – the 'CAS' system – uses three morphological parameters: the image concentration (C), the asymmetry (A), and image clumpiness (S) (Conselice, 2003).

The asymmetry parameter is evaluated by rotating the image of the galaxy 180 degrees and subtracting it from the unrotated image, and comparing the sum of these residuals to the original image. The asymmetry is then the square root of the sum of the squares of the residuals (the pixel-by-pixel difference between the original and subtracted images) normalized by the twice the sum of the squares of the values in the original image (Conselice et al., 2000b). Analysis shows that as long as the a physical scale of 0.5 h_{75}^{-1} kpc can be resolved, the asymmetry can be reliably determined. However, this is subject to (1) galaxies being physically large, and (2) the imaging survey being deep enough to detect the principal components of the galaxy, such as a (possibly faint) disk.

The concentration parameter is based on a scaling of the light distribution using a value 1.5 times the Petrosian (1976) radius at which $\eta \equiv I(r)/\langle I(<r)\rangle = 0.2$, where η is the ratio of the surface brightness at r to the average surface brightness interior to r. At this distance, 100% of the light is interior to this Petrosian radius. In this scheme, $r_{50\%}$ is roughly the half-light radius. The concentration index is defined as $C \equiv 5 \times \log_{10}(r_{80\%}/r_{20\%})$ (Conselice, 2003), where $r_{80\%}$ and $r_{20\%}$ are the radii at which 80% and 20% of the light is respectively interior. This parameter is sensitive to the shape of the distribution of light, not the smallness of its physical size. The de Vaucouleurs $r^{1/4}$ profile has C = 5.2, while the exponential profile has C = 2.7 (Bershady et al., 2000).

The clumpiness parameter is designed to find the high-frequency spatial variations of star-forming regions. It does this by convolving the galaxy image with a spatial filter (of size $\sim 0.4 r_{20\%}$) to wash out high-spatial frequency components, then subtracting the result from the unfiltered image. Pixels with negative values in the residual image are set to zero. The clumpiness, S, is defined as ten times the sum over pixels of the ratio of the value of residual image minus the background pixel values in empty sky to that of the original image. They do not consider the inner part of the galaxy. This leads to E-galaxy clumpiness near zero, for late-type disk galaxies $S \sim 0.4$, irregulars $S \sim 0.45$, and starbursts, $S \sim 0.7$.

The highest asymmetries ($A > 0.35$) are produced by obvious mergers (Conselice et al., 2000b,a). "Early" galaxies are well separated by asymmetries $A \lesssim 0.05$. Mid- and late-type galaxies are distinguished by a combination of asymmetry and concentration parameters. A strong concentration index C (values ~ 3 to 4) usually suggests a dominant bulge (or E) system, and a low light concentration ($C \sim 2$ to 3) goes with a diffuse-appearing galaxy of fairly gradual radial surface brightness decline. Of the more highly concentrated galaxies, early and late galaxies must be distinguished with the asymmetry index. The clumpiness S index may not be very useful for independent morphological data at the highest redshifts without a significant advance in spatial resolution. Of the remaining two parameters, the asymmetry is arguably the most important.

Generally, as one probes to higher redshift, and the systematic galactic star-formation rate increases, galaxies become bluer and generally more asymmetric (due to localized SF regions). A recent Conselice et al. (2004) paper reports on deep ACS observations of faint galaxies, and compiles photo-z and morphological analysis of 2354 objects, mostly galaxies with $M_{\mathrm{B}} < -19$, found between $z = 0.2$ and $z = 2.0$. Their average concentration index is $C = 2.7 \pm 0.5$ and the asymmetry index is $A = 0.26 \pm 0.1$ with *outliers*, while local galaxies have mean concentrations of about 3.5 ± 1.5, and asymmetries of about 0.1 ± 0.1. More recent work using 1231 $I < 27$ galaxies from the HDF North and South (Conselice et al., 2005), probes to almost $z = 4$. The scatter plots in Fig. 8 show the trend in the A versus C indices with redshift. As one moves from lower redshift subsets ($0.2 \leq z \leq 0.7$), to high redshifts ($2.0 \leq z \leq 4$), note the decrease in the concentration index from an average near $C \sim 3$ to $C \sim 2.4$. The extremely low concentration indices (> 1-σ from the average) at $C < 2.2$ are called luminous diffuse objects (LDOs); Conselice et al. 2004). Sample LDOs at progressively larger redshifts are shown in Fig. 7.

The galaxies with high asymmetries ($A > 0.35$) are usually those undergoing tidal disturbances – likely mergers or post-merger views, as per the preceding van den Bergh images (Figures 5 and 6). Their abbreviated code for luminous asymmetric objects is naturally LAOs. Perhaps 2% of the ACS galaxies with good photometry show simultaneously both indices in the atypical range (i.e., large A, and low C). But as far as individual outliers in one dimension ($A > 0.35$, say) the LAOs are moderately common (perhaps $\sim 20\%$ of the full sample).

This asymmetry cut allowed Conselice (2003) to impersonally identify galaxies undergoing tidal disturbances. Simulations show that the redshift alone could not displace nearby galaxies to these atypical portions of (for example) C – A space.

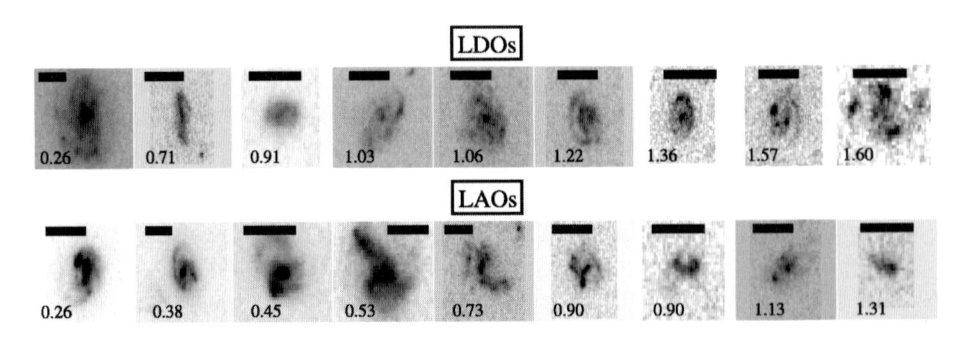

FIGURE 7: Various images of LDOs (luminous and diffuse) and LAOs (luminous and asymmetric) galaxies. Bar shows approximately $0''.5$. Redshifts, mainly photo-zs, are given by the numbers in the lower left corner of images. These redshifts are thought to be reliable at the moderate distances in question here. The figure is from Conselice et al. (2004).

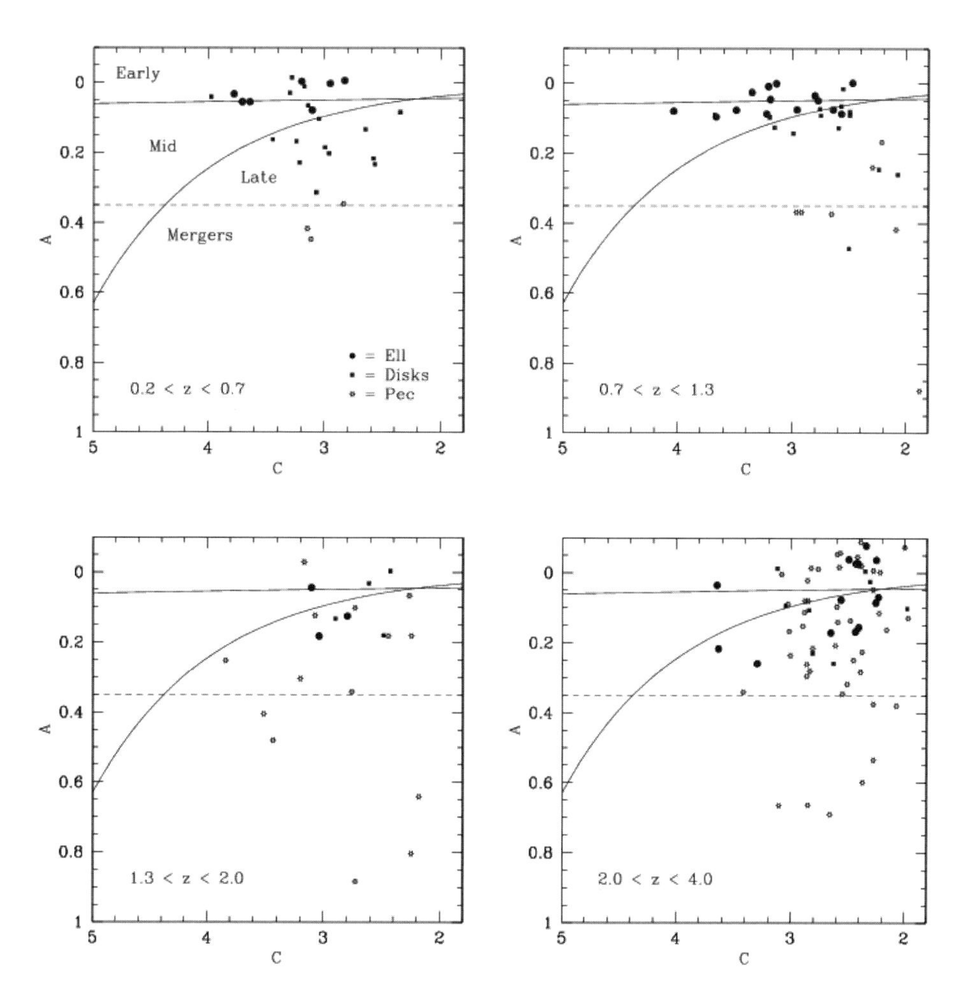

FIGURE 8: Scatter plot of asymmetry versus concentration indices as a function of redshift for elliptical, spiral, and peculiar galaxies. The solid line and curve separate regions where early, mid, and late-type galaxies are found locally, as shown in the first sub-panel. Notice that mergers, which appear below the dashed line, become relatively frequent at $z > 1.3$. The diagram is from Conselice et al. (2005).

The LAOs and LDOs are abundant in their co-moving volume density between $z = 1$ and 2. For the LDOs their density in the past was similar to that of *local* bright disk systems. That represents a lot of ancillary SF early on.

The LDOs and LAOs apparently host $\geq 1/3$ of the SF activity in galaxies in an active cosmic interval; many also "look like accidents", as well as showing the impersonal morphological peculiarities in the indices we discussed. We note that in Fig. 8 from Conselice et al. (2005), at $z \sim 3$, at least half the galaxies seem atypical in their asymmetry indices (in terms of the present) – the mergers that occur at $z \simeq 3$ could be the dominant evolutionary event in the youth of galaxies and their smaller "cousins", the pre-galactic fragments (whether they have formed stars yet or not).

It is clear that this classification system will be useful in categorizing the galaxies detected in the immense data-sets now being produced (§1.7). However, it will be important to use it carefully. We note that morphological analysis of high redshift objects is not only made difficult by the spatial resolution problems, but also by the photometric band-shifting that requires we observe galaxies at ever-longer wavelengths for ever-higher redshifts. In addition, the detection of faint (LSB) features is jeopardized by the $(1+z)^4$ cosmological surface brightness dimming due to the effects of redshift on the received flux. We must be mindful that our comparisons with low-redshift objects suffer when the surface brightness limits for high-z objects are not as deep in the rest frame as more thoroughly studied low-redshift galaxies. The effects of the declining depth of surface brightness with redshift can be studied at low-redshift by morphological analysis of images taken with a range of surface brightness limits to study the change in the CAS indexes. Conselice (2003) finds that such surface brightness and resolution effects are moderate for local galaxies when they are rebinned to simulate progressively higher redshifts. It is found that the extracted concentration indices stay constant, or increase at the 5–10% level, as one simulates the galaxy at redshifts $z > 1$. However, for redshifts $z \gtrsim 1$, the asymmetry indices are found to decline by values approaching the average asymmetry at low redshift – hence making the results uncertain. A similar situation is found for the clumpiness index S. Thus, the concentration index will probably continue to be the most useful CAS index at high redshift, at least until the next generation of giant telescopes with adaptive optics is built (see Chapter 6).

3.4 Explaining "Building Block" Galaxies at High Redshift

A small diversion into thoughts about what the smallest and earliest star forming objects might be iin order. We have seen that the half-light diameters of galaxies seen at high redshift get progressively smaller with increasing redshift (Fig. 4), to the point where they begin to be called components, fragments, or sub-galactic clumps. They appear to represent the smallest gravitationally bound systems capable of inducing their own star-formation. Accordingly, these might be referred to as "building blocks". Because these systems are morphologically distinct from larger galaxies, we suggest that the formation mechanism for such small, compact galaxies may be fundamentally different from that of larger galaxies. Lyman alpha galaxies – galaxies whose spectra often detect only the Lyα line in emission – are a good example of these systems, and will be discussed in the next chapter. The compact nature of lower-redshift ($z \sim 0.2$) specimens is lost when the stellar velocity dispersions rise above about 50 km s^{-1} (Guzman et al., 1996). We speculate that other modes of SF and galaxy growth, such as shock-induced SF or the merger of other systems, may in time become dominant influences in galaxy growth, producing the transformation from a compact to a more diffuse morphology over reasonable cosmic epochs.

These systems are not fragments in the normal sense of the word since they are most often seen in isolation. They are not truly sub-galactic since they are galaxies. We could call them *primeval* galaxies, but this refers to the stage of their origination rather than their morphology, and may therefore miss the mark. Morphologically, they are a type of galaxy most often compared to the local H II galaxies and the somewhat more distant compact narrow emission-line galaxies Koo et al. (1994). If these systems are seen undergoing their first burst of SF, the word "primeval" would seem to fit, though a residual reluctance might stem from the preconception that "primeval" must refer only to the first galaxies.

Accepting the view that these are newly forming galaxies, would require the abandonment of the assumption of the coevality of galaxies; galaxies of a similar size and morphology have not necessarily formed in a lockstep march from small to large systems. The coevality of galaxies can be seen to be a naive application of the concept of the hierarchical assembly of structure. Applied to a universe that is uniform on large scales, small things form first, large things form later. However, applied to a non-uniform universe, we must acknowledge that the most rapid morphological evolution of systems occurs in the denser agglomerations of matter. Galaxies in low-density environments get few mergers, and are basically supply-limited. Though galaxies appear to be non-coeval, hierarchical assembly is not contradicted – it just proceeds at a rate dependent on the local comoving density.

How do we explain the very small sizes of these first-generation galaxies in terms of galaxy-formation theory? The Jeans length, which is the scale at which homogeneous baryonic clouds become unstable to collapse, is the standard starting point for modeling gravitational collapse. The Jeans length evolves as the cosmic scale-length $a(t)$ (i.e., as $(1+z)$), as can be seen by noting the trend of dropping temperature T ($T \propto a^{-1}$) and density ρ ($\rho \propto a^{-3}$) with increasing $a(t)$ and with time;

$$\lambda_j = (\pi k T/(G\rho_{\rm b} m_{\rm p}))^{1/2}, \tag{1}$$

where $\rho_{\rm b}$ is the baryon density, $m_{\rm p}$ is the mass of a proton, k is Boltzmann's constant, and T is the temperature. The Jeans *mass*, however, which is the product of baryonic density times the Jeans length cubed, is independent of time and redshift in a homogeneous universe,

$$M_{\rm j} \simeq \frac{\pi}{6}\rho_{\rm b}\lambda_{\rm j}^3 \sim 10^5(\Omega_{\rm b}{\rm h}^2)^{-1/2} \simeq 7.6\times10^5\ {\rm M}_\odot. \tag{2}$$

Globular cluster-size fragments of galaxies ($M \lesssim 10^6$ M$_\odot$) are possible under this model. Are the small first-generation galaxies relicts of this form of structure formation? This is somewhat doubtful, as we will presently see, since a spherical monolithic collapse forms stellar systems that are less compact than our observed "building blocks". Though these compact systems may thus begin with baryonic masses of 10^5 to 10^6 M$_\odot$, their formation epoch is roughly that at which they are seen since the specific SFR is high, and they cannot remain seen in their diminutive size for long. The effects of gravity then lead to mergers with other systems, gradually building galaxy mass, and gradually contributing to a large scale structure (LSS).

Jeans mass calculations are highly idealized and may not reflect conditions in the proto-galactic clump. Another cloud model exists – that of the isothermal cloud in a cosmic setting. Homogeneous versus isothermal cloud collapse and star formation were studied with numerical simulations (Murray & Lin, 1992). The results showed that the monolithic collapse of a uniform cloud produces a larger galaxy than that produced by inside-out collapse of an isothermal cloud. The inside-out galaxy formation model was used to simulate half-light radii of simulated galaxies (Cayón et al., 1996; Bouwens et al., 1997), and resulted in a good match with high-z observations when an open model is used. Manning et al. (2000) found two possible protogalactic units at $z \sim 3$, seen only in Lyα emission, and with half-light radii of $r \lesssim 1\ \mathrm{h}_{70}^{-1}$ kpc. Such small stellar systems make a good prototype of inside-out star formation in a Bonnor–Ebert sphere (Shu, 1977). A Bonnor–Ebert sphere results from enveloping an isothermal gas sphere in a hot medium which applies an external pressure on the sphere. This pressure, if high enough, causes the sphere to collapse from the inside-out. The resulting star-forming regions are quite compact; modeling work shows radii are about 1 kpc (Lin & Murray, 1992; Cayón et al., 1996), consistent with the observations of Manning et al., and "building block" galaxies generally. Figure 4 shows that average half-light radii for galaxies approach values $r_{1/2} = 1$ kpc at high redshift when interpreted with the flat-Lambda cosmological model. This suggests that, at high redshift, the galaxy LF is dominated at the faint end by these compact star-formers.

Though both models are simplifications, they provide the starting point for understanding the earliest stages of galaxy formation. The small angular diameters of high-redshift galaxies suggest a precursor for our "primevals" that is closer to an Bonnor–Ebert (isothermal) sphere.

3.5 One Small Problem – Resolving Galaxy Bars

We wish to note one small specific morphology problem: are spiral galaxy *bars* visible at moderate to large redshifts? There is an oft-repeated claim that spirals with strong (easily visible to the eye or perhaps through an impersonal symmetry index) bars make up a noticeably smaller fraction of the disk population at $z \geq 0.7$ (van den Bergh et al., 2000, 2002; Abraham et al., 1999).

If this suggestion were to be determined robust, and there is a real evolutionary difference between $z = 0$ and $z \sim 0.7$ spirals, then it would have significance in the astrophysics of disk formation. However, as Bunker et al. (2000) point out, the effects of large morphological K-corrections (viewing the nearby and distant spirals at differing rest-frame wavelengths) make the case for physical evolution much more marginal. Bars are usually populated by an older, cool stellar population, as shown in Fig. 9, and thus are prominent at longer (IR) wavelengths. But at $z > 0.7$, say, we often observe the rest ultraviolet where the old stars can be

FIGURE 9: The nearby spiral NGC 7424, taken with VIMOS/VLT at the European Southern Observatory. Note the prominent bar; the redder colors of the bar (see color section) indicates an older population of stars populate the bar. Such features are difficult to detect at high redshift because our optical image then views the rest UV, and old stars emit very little UV. Imaging in IR bands may be of use in detecting bars at high redshift, as per the work of Bunker et al. (2000). (Reproduced in colour in the colour section.)

dominated by any residual SF, thus often breaking any morphologically smooth pattern like a bar.

So a direct exercise to resolve the problem was attempted by Bunker et al. (2000); they used near-IR NICMOS J and H images of the Hubble Deep Field North (HDF), observed at $\lambda = 1.2$ μm and 1.6 μm, so that galaxies at $z = 0.7$ (1.0) are seen in their proper wavelengths of 9400 Å (8000 Å) and 7060 Å (6000 Å). These rest-wavelengths strongly bias the images toward the contributions of cool stars (types later than the Sun) and thus would favor bar visibility. Examination of the NICMOS HDF images do reveal some bars in the IR which turn out to be undetected in the shorter-wave WFPC2 HDF images. More recent work (Sheth et al., 2003) confirms this, concluding there is no significant evidence supporting a decrease in the barred spiral galaxy population beyond $z = 0.7$. Thus we favor the view of a merely *apparent* difference in the barred fraction of distant S galaxies, rather than any true evolutionary difference.

This situation highlights a general problem of morphological analysis at high redshifts: one must observe in bands which are redshifted relative to the bands in which features are seen in local galaxies.

3.6 Is there Evolution in the Stellar Content of E Galaxies?

Nearby elliptical galaxies have a dominant "cool" stellar content and a color–magnitude array which, at least superficially, looks similar to that of an old galactic globular cluster. But the look-back time to redshifts $z \simeq 1$ to 1.5 is $\gtrsim 8$ Gyr, so that more-distant giant E galaxies, if they have the character of nearby dormant systems, may well be hard to recognize. That is because at the present epoch, the E galaxies are all "old and dead"; their hottest stars generally being near the main sequence at a turnoff spectral type in the late F stars or even G0. The age of such stars approaches 5 Gyr. At higher redshifts, we anticipate a "bluing" of the turnoff and a slightly- to moderately-bluer overall galaxy. This reduces the redness of the system, and could constitute a substantial effect by $z \gtrsim 1.6$.

A few investigators have pushed the study of E galaxies beyond $z = 1$. At this distance, they are still found in groups and clusters, and they are occasionally identified with radio sources. Using such non-traditional search aids as radio-loudness are justified because we may wish to utilize all parameters as clues to the assembly of massive galaxies or massive spheroids.

The two types of evolution we contrast are first, a single monolithic collapse, with infall from the immediate surroundings, with little SF beyond the initial burst. The following "dormant" stage would likely be observable over $0 \leq z \leq 2$. The alternative model of galaxy-building, perhaps relevant to the structure of giant E systems and other galaxies, is the hierarchical merging "tree" (Wechsler et al., 2002). Here, mergers are accompanied by SF bursts, but eventually build up a relaxed-looking galaxy baryonic mass (stars and gas).

The timing and size of merging events can affect the resultant galaxy color and spectral character, with very noticable consequences. The most-sensitive portion of the spectrum for detection of recent SF in a predominantly old galaxy is the rest frame range in the near-UV, $\lambda\lambda_0 2000$–3000 Å. This situation can be detected by quantitative spectroscopy at (observed) optical wavelengths, with the practical disadvantage being the observed faintness of the galaxy with negligible SF bursts.

Spinrad et al. (1997) chose candidate Es by choosing symmetric, red galaxies that may be weak radio sources or companion systems to radio-loud galaxies. Their spectroscopy was focused upon the continuum slope near rest 2700 Å ($\pm$300 Å) and discontinuities at $\lambda\lambda_0$ 2900 and 2640 Å, caused by Mg I and Mg II absorption and hosts of Fe II blends, especially just below the discontinuities at 2900 and 2640 Å. These breaks are strong in late F/early G stars, and weak in hotter stars. They yield a robust population age, assuming a metallicity near solar.

To test for evolutionary change, spectroscopy by Spinrad et al. (1997) found one faint E-type system, observed in the rest-frame UV at $z \simeq 1.55$. An analysis using the Bruzual A. (2001); Bruzual A. & Charlot (1993); Bruzual A. (1983) stellar evolution models was used to interpret the spectrum. It was found to be consistent with a main sequence turnoff at an F6 V star. At this redshift, the Universe was about 4 Gyr old. A main-sequence turn-off at F6 stars suggests that the last significant burst of star formation occurred perhaps 2–3 Gyr earlier, at a redshift $z \gtrsim 4$. This is an epoch at which hierarchical assembly models predict merging as an active process for proto-Es. What characterizes Es (and S0s) is *passive evolution*, in which a stellar population only evolves by the well-known evolution of existing stars; no new star formation takes place. The main effect is a slow decrease in the luminosity of the main sequence turn-off stars, and the G and K subgiants. If the evolution since a burst occurring ~3 Gyr prior to the epoch of observation is passive, then fits to stellar evolution models with no post-event SF, or with an exponential decay in SF, yield robust stellar histories; in particular, 3 Gyr age models do fit the break amplitudes above (Spinrad et al., 1997; Dunlop et al., 1996; Nolan et al., 2001). This trend if extended by analysis of other red E systems (e.g., McCarthy, 2004) – would favor early major SF, and not much since $z \sim 3$ or even, perhaps, $z \sim 5$. For such systems, an intrusion of younger stars would be easily noticed. So here a major property – a SF-rejuvenated UV spectrum – is constrained by E galaxies that appear to be passively evolving their old stars and nothing more.

Some recent observations may suggest an even more rapid end to SF than previously thought in ancient, distant ellipticals. Through various color-selection techniques, research groups are now finding proto-elliptical galaxies at redshifts in excess of 2. Daddi et al. (2004), Cimatti et al. (2004), and Schreiber et al. (2004) find large numbers of $M_* \sim 10^{11}$ $M_\odot$ galaxies at $1.6 \lesssim z \lesssim 2.5$ (i.e., comparable numbers to that of LBGs observed by Steidel et al. 1999), with a number density on order $10^{-4}\,\mathrm{Mpc}^{-3}$. This is a bit of a surprise because semi-analytical modeling under a ΛCDM model (e.g., Kaviani et al., 2003) generally project number densities smaller by an order of magnitude than observed. Most of these high-z objects are rapid star-formers at the time of observation, but a significant number (~25–30%) are passively evolving (Daddi et al., 2004). Thus, mass-assembly would have to have been significantly more efficient at $z \gtrsim 3$ than most models of early evolution allow, with red galaxy comoving number densities about 10 times greater than predicted. This discovery is too recent at this writing to report a probable response by the modeling community.

An alternate (but similar) study can be focused on photometric observations; McCarthy (2004) identifies a red-color envelope (maximum redness) for E galaxies within the redshift range $0.5 \leq z \leq 1.6$. The trend in color implies only passive evolution for some of the E galaxies during the last few gigayears, suggesting that the last epoch of SF was at $z_{\rm form} \gtrsim 5.0$ (McCarthy, 2004), roughly 11 Gyr ago. This would suggest a minimalistic SF history for much of the lifetime of giant E-systems – a rule that applies to both field and cluster E galaxies.

Large samples can be used to monitor the evolution of early-type galaxies at lower redshifts. Bell et al. (2004) used deep multi-band photometry in 0.78 deg^2 and photometric redshifts to look at the redshift dependency of $m_{\rm R} \lesssim 24$ mag galaxy colors. They find distributions that are bimodal in color, and vary with redshift. The fraction of galaxies that are blue has declined since $z \sim 1$, and the mass of the red-population has nearly doubled. Does this mean SF in early-type systems? Bell et al. suggest that galaxy merging, and the truncation of SF in blue galaxies, are the likely causes for the growth in red-sequence mass; no significant SF is required to explain the change. A somewhat related investigation (Drory et al., 2004) of K–selected galaxies over $0.4 \leq z \leq 1.2$ produced similar results. Since the K-band luminosity correlates well with stellar mass, Drory et al. were able to measure the comoving stellar mass density in massive ($10^{10} \leq M \leq 10^{12}$ $M_{\odot}$) systems since $z \sim 1$, detecting a doubling of " passive" mass in luminous systems over this interval. They conclude that massive galaxy growth is dominated by accretion and merging rather than by SF, a conclusion which harmonizes with that of Bell et al.

3.7 Galaxy Clusters – Now, and as they were at Moderate Redshift

Rich clusters of galaxies are the largest collapsed structures we can observe, and perhaps model successfully. Clusters can be studied as a group by calculating their integrated X-ray or optical luminosity functions, and their galaxy type distributions. One may also gain insight into their mass and degree of relaxation by modeling individuals as large collections of test particles in cosmic simulations. The abundance of massive clusters can provide a constraint on the matter-content of the Universe.

We observe and discover distant clusters by means of deep optical imaging (Stern et al., 2003; Stanford et al., 1997; Stanford et al., 2002) and X-ray emission (Rosati et al., 2002b; Ebeling et al., 2000; Rosati et al., 1999, and references therein) and (a few) by the presence of a massive radio galaxy in a luminous E-galaxy host. An excellent review is available (Rosati et al., 2002a).

Soon, clusters will be detected in large numbers by a new generation of Sunyaev–Zel'dovich effect (SZE) microwave detectors (Sunyaev & Zeldovich, 1969). The SZE is the spectral distortion produced by inverse Compton scattering of CMB photons in the ionized cluster medium (refer also to Figure 8 in Chapter 1). The fraction of photons energetically shifted by their path through the cluster is a measure of the gas temperature, and hence the mass of the cluster. Figure 10 shows the increase in the CMB Planck spectrum temperature from inverse Compton scattering in the interferometric imaging of the SZE with the Owens Valley Radio Observatory (OVRO) (Joy et al., 2001).

Many ambitious instruments are being planned to study the SZE. The upcoming *Planck Surveyor* satellite should provide an all-sky survey of the CMB. The design philosophy is to have very broad frequency coverage by using both a low-frequency instrument, utilizing HEMTs (high electron-mobility transistors in the frequency range 30–100 GHz), and a high-frequency instrument (100–850 GHz) using bolometers. This duality will provide maximum discrimination between the

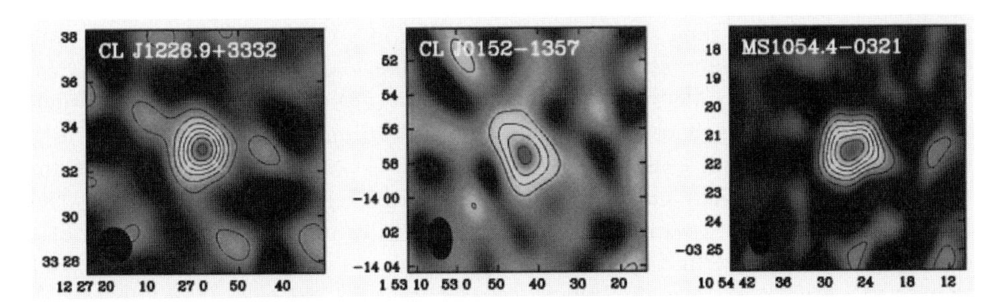

FIGURE 10: Synthesized images of the SZ effect for three cluster fields. Contours represent multiples of 1.5-σ, where σ is the RMS flux. The data can be used to fit a cluster electron density model to the SZE decrement (usually an isothermal β model) to help derive total masses and cluster temperatures of the clusters. Derived masses for each of these clusters are $\gtrsim 2 \times 10^{14}$ $M_{\odot}$within 340 h_{100}^{-1} $M_{\odot}$. Figure from Joy et al. (2001). (Reproduced in colour in the colour section)

foregrounds and the CMB. It will be able to measure $\sim 10^4$ clusters within redshift 1.0 (Carlstrom et al., 2002). Inevitably, the SZE data will be more useful when constrained by X-ray, weak and strong lensing, and gas and galaxy velocity dispersion measurements. One of the expected results of this effort is a well constrained mass function for clusters as a function of redshift.

Cosmological parameters may also be constrained by a cluster study using the SZE, though it is difficult to tell how well this can be done until the systematic uncertainties are understood. Included in these systematics are the spatial and temporal variations in the emission from our atmosphere and the surrounding ground, as well as gain instabilities inherent to the detector system used. There are also sources of astronomical confusion such as the CMB anisotropy itself, point radio sources, dust emission from certain galaxies and gravitational lensing by foreground material. A "full-court press" using independent cluster information (see above), utilizing the spectral resolution of single-dish receivers and the imaging powers of interferometric systems, should enable a good understanding of many systematic concerns.

Our main task here is to examine the evidence in hand for the evolution of the observable constituents of these clusters – their galaxies (mainly, but not completely E/S0 systems) and their hot gas halos (the global X-ray emission source of the intracluster medium).

We'd like to count the rich (massive) clusters by using an efficient method to locate the clusters over a wide z-range, to estimate their masses, and derive a method to robustly handle the survey volume in a way that accounts for biases. At this epoch, detecting their X-ray luminosity is the best method to accomplish these goals.

The X-ray luminosity uniquely determines the cluster-selection function. The X-rays arise through the equipartition of energy; the gas temperature is closely linked to the virial velocity of the component galaxies – itself a function of the cluster gravitational potential – and thermal emission from the hot gas is large enough to emit X-rays. The instruments needed for the wide-angle X-ray surveys

(ROSAT) and more physical studies (ASCA, Beppo-SAX, and Chandra) are now (or were recently) adequately functional for our first explorations at keV energies.

X-ray selection is a sound method to illustrate physically bound systems such as clusters of galaxies; there the hot Intracluster Medium (ICM) strongly suggests a deep cluster potential well. The gas is assumed to be in dynamical equilibrium with the stellar galaxies and the "ever-present" dark matter. Consistently, the X-ray luminosity is well-correlated with overall cluster mass – $L_X \propto M^{1.8}$ (Rosati et al., 2002a). So X-ray initiated searches at higher z for rich clusters are feasible, though the cosmic surface brightness dimming $\propto (1+z)^4$ takes a heavy toll at high redshift. The ROSAT sky-survey, and specifically pointed searches, yielded over 1000 candidate galaxy X-ray clusters within a redshift of 0.5 (Rosati et al., 2002a). An X-ray LF of ROSAT clusters are shown in Fig. 11 (Böhringer et al., 2002). It displays a double power-law structure, with a break at about 5 keV.

Of course, the local density of rich clusters must be determined before any evolution from substantial redshift can be robustly determined.

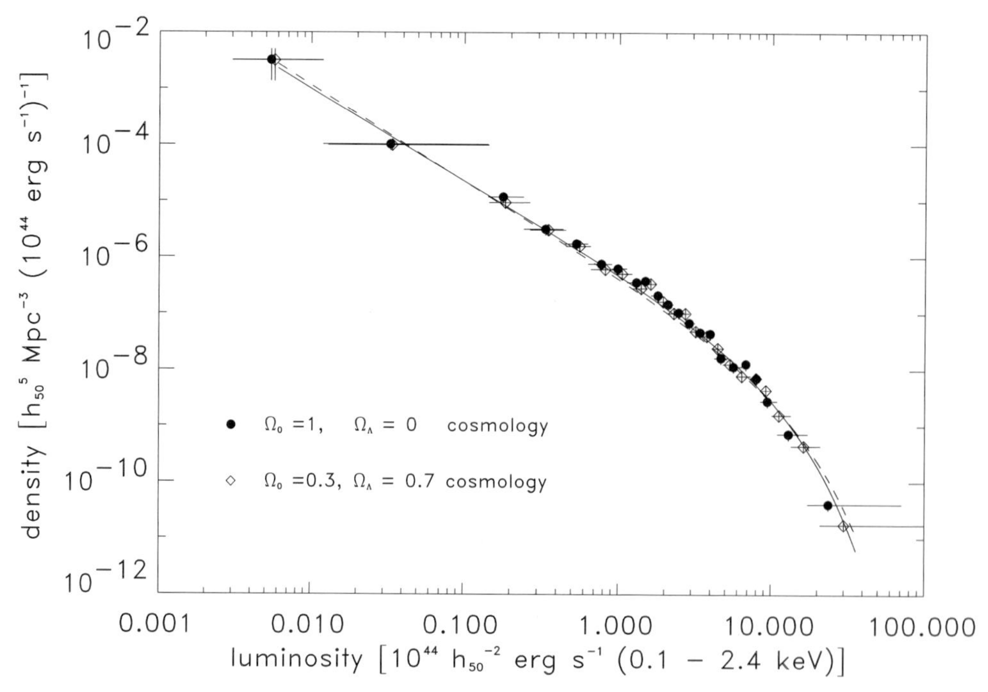

FIGURE 11: The REFLEX X-ray luminosity function derived for an Einstein–de Sitter cosmology, and for a flat $\Lambda = 0.7$ cosmology. Observationally, there is very little difference between the two. The XLF has been modeled as a double power-law. Figure courtesy of Böhringer et al. (2002).

Rosati et al. (2002a) compare cluster X-ray luminosity functions (XLF) for a few different studies, ranging from the Einstein Medium Sensitivity Survey (at $z \approx 0.1$) to the ROSAT ESO Flux-Limited X-ray (REFLEX) sample, and found no suggestion of significant variation with redshift from $z \simeq 0$ to $z \simeq 0.5$. Though the lower luminosity side of the cluster XLF shows no significant luminosity evolution within $z = 1$, there does appear to be a moderate 'shortage' of distant, luminous clusters with $L_X > 10^{44}\,\mathrm{erg\,s^{-1}}$.

The relatively new X-ray capabilities of Chandra for locating distant X-ray clusters has led to the detection a few clusters beyond $z = 1$ (see Stanford et al., 2001). But these new data don't change our view of just a slight increase in the XLF on the bright end with cosmic epoch (e.g., a slightly stronger luminous side at the present time, $z \simeq 0.1$ or so).

Thus, our view of rich X-ray clusters is one of a constant population to $z \sim 1.0$, perhaps with a modest rarity in the most luminous (massive) clusters at $z > 0.7$. Such a rarity of the richest clusters may simply reflect the insufficient amount of time to assemble or virialize a rich cluster; for an initial formation epoch of $z \approx 2$, rich clusters seen at $z = 1$ have only ~ 2.5 Gyr to virialize. This may not be an adequate amount of time to smooth and deepen the collective cluster potential well, as crossing times for the central 2 Mpc are about 2.4 Gyr for clusters with initial radial velocity dispersions on the order of 500 $\mathrm{km\,s^{-1}}$ (Bahcall, 1977).

3.8 Global Star-formation: Changes from $z \leq 1$ to Here and Now at $z = 0$

In many individual galaxies the stellar content, SFR and their evolution may cover a wide astrophysical range – perhaps stochastically over a set of contemporaneous galaxies. However, globally we can begin to feel confident about some of the systematic trends in the galaxy SF (as a function of z).

In this section we deal with the near-past epoch of global SF density integrated over substantial cosmic volumes and including observations of many individual galaxies. For the moment we shall consider the redshift range from $z = 0$ to $z \simeq 1$, covering a look-back time span of about 8 Gyr with the flat Λ cosmology and $H_0 = 70\ \mathrm{km\,s^{-1}\,Mpc^{-1}}$. The last ~8 Gyr years has shown a dramatic change in the global SFR density as well as in galaxy morphology (discussed earlier in §3.2).

Galaxies with disturbed morphologies, especially the blue ones, seem much more numerous on moderately deep images and spectra, including the intermediate ($z \sim 1$) redshift territory. Lilly et al. (1995) (the Canada-France Redshift survey; CFRS) provide a good starting point for the systematic evaluation of the evolution of field galaxies and their luminosity function out to a redshift of about unity.

The CFRS team obtained redshifts and photometry for 591 galaxies (with reliable redshifts), with an I-band limiting magnitude of $I_{\mathrm{AB}} = 22.5$ mag. Using the near-IR as a selection wavelength was convenient and "safer" than previous galaxy surveys because it captures the rest B and V-band range which comes from longer-lived stellar components. The CFRS used their $(V - I)$ mags and colors, which straddle the rest-frame B for $0.2 < z < 0.9$, just the redshift range of import for this stated goal.

Since the epochal Lilly et al. (1995) paper was published, several research teams have tried to reproduce its main result – the great abundance of star-forming, blue galaxies (Sb, Sc, Sd, Irr) at intermediate redshifts ($z > 0.5$). The SFR density is found to increase rapidly with increasing redshift from $z \sim 0$ to $z \sim 1$ by the large factor of 10–20 times (Flores et al., 1999; Glazebrook et al., 1999; Rosa-González et al., 2002; Madau et al., 1996; Fried et al., 2001). The tendency toward a more vigorous SF rate in galaxies beyond $z \sim 0.5$ is noticeable in both intrinsically bright and intrinsically faint spirals.

Fortunately, several techniques and several wave-bands, mainly emphasizing continuum observations, have been utilized to determine the SFR out to redshift one. Their agreement is heartening! The galaxy emission lines can also help in SFR determination.

Specifically, the Hα spectroscopy of Glazebrook et al. (1999), and the Flores et al. (1999) use of the Infrared Space Observer (ISO) and 15 μm photometry, represent rather different techniques and selections. They still matched well to the differential SFR density differences between old/new comparisons from $z \sim 0$ to $z \sim 1$. They also confirmed the SF rate difference from the earlier Madau et al. (1996) and Lilly et al. (1995) papers. An interested reader can also see how the various wave-band calibrations agree in Rosa-González et al. (2002). We'll discuss the evolution of the gaseous emission-line fluxes in normal galaxies over $0 \leq z \leq 1.0$ in the next section of this chapter.

A simple astrophysical accounting and interpretation of the galactic SFR and its space-density evades our theoretical consensus, but is likely to be related to the excess B-counts of blue galaxies at moderately deep levels by Cohen et al. (2003).

One interpretive paper by Kolatt et al. (1999) may explain the abundance of these blue galaxies. They asked the question: Why the relative paucity of local actively star-forming galaxies (or halos with large central densities of baryons) compared to $z \sim 1$ systems? Under the hypothesis that major mergers are the principal cause of starbursts, they utilize N-body simulations and a hierarchical halo-finder in a flat Λ cosmology to estimate the galaxy collision rate. Their findings suggest that starbursts associated with mergers of relatively low-mass halos may dominate the high-z SFR. Mergers are assumed to cause an enhanced luminosity for $\sim$ 100 Myr. At lower redshifts ($z < 2$), observational data (Patton et al., 2002) has suggested that the merger frequency declines sharply with time (increases with redshift) as approximately $(1+z)^{2.3}$ out to $z = 1.5$ or 2.

3.9 Emission Lines as a Star-formation Proxy, out to $z = 1$

Besides the galaxy UV-blue continua that help to determine the more local SFR, we can utilize emission lines as a "shortcut" method to estimate to the luminosity density of the less local Universe. Here we note recent attempts to use a "solid" recombination line such as H_α, and then, as astronomers push to greater distance, we must choose progressively shorter rest-wavelength emission lines, like the [O II] doublet at $\lambda_0 = 3727$, as a surrogate for H_α and (less directly) the UV continuum from hot stars.

This work has been dominated of late by a few groups; in particular we concentrate on the [OII] line research of Hogg et al. (1998), and the effort concentrating on several lines, including [OII] and Hα, by Hippelein et al. (2003).

The [OII] emission of a galaxy depends upon its recent SF history, as [O II] emission occurs in H II regions around OB associations. The luminosity of the line has been calibrated by Kennicutt (1992) against Hα and against the SFR determined from the galaxy continuum. The stochastic nature of dust extinction along the multiple sight-lines to the galaxy and around to individual H II regions, poses problems for the calculation of the internal dust distribution. Thus the [OII] line correlation with the SFR is noisy.

The Hogg *et al.* paper discussed the equivalent widths (EW; roughly, the line flux relative to that of the continuum) in 375 faint galaxy spectra taken for a redshift survey. They then calculate the [O II] luminosity function (in its galactic rest-frame), though the work may be a bit unclear about explicit corrections that were made for wavelength-dependent dust extinction.

The conclusion of the Hogg team is that the luminosity density of the (OII) 3727 emission line is a factor of 10 higher near $z = 1$ than at the present epoch. Presumably that dramatic change mirrors the SFR density of the Universe; it is in accord with Madau et al. (1996) and Madau et al. (1998) who use broad-band photometry of galaxies in the HDF and ground-based surveys.

The second independent reference on the emission-line topic by Hippelein et al. (2003) provides an independent study of the emission-line galaxies. The survey is both spatially wide (500 $\square'$), and is fairly deep (to a limiting flux of 3×10^{-17} erg s^{-1}) for the Hα, [OIII], and/or [OII] lines. Having access to all three relatively strong galaxy emission features is an advantage since the strong emission-line redshift windows will now allow a wider range in redshift for spectroscopic follow-ups. This Calar Alto Deep Imaging Survey (CADIS) group (named after the observatory of the same name) has measured the SFR over the redshift range $0.24 < z < 1.21$, with Hα providing most of the small-z redshifts and line fluxes and [O II] λ_0 3727 covering the high-z end.

Extinction corrections were performed and a luminosity function derived. Then to determine the SFR from the corrected emission line luminosities we use the conversion factors listed in Table 1, where the star formation rate is

$$\mathrm{SFR}(\,\mathrm{M}_{\odot}\mathrm{yr}^{-1}) = \mathcal{K}\,\mathrm{L}_{\mathrm{line}}\,(\mathrm{erg\,s}^{-1}), \tag{3}$$

from Kennicutt et al. (1994), where $\mathrm{L}_{\mathrm{line}}$ is the line-luminosity. Other conversion factors are in place for the other emission lines.

The extinction-corrected UV continuum flux density at 2800 Å as a function of the various emission-line luminosities L (see Fig. 12) shows a good correlation among the various emission lines, but has significant scatter. This suggests that these line luminosities can be good estimators of the SFR to $z \gtrsim 1.0$. Hippelein et al. (2003) has employed this technique to calculate the SFR density. Figure 13 shows the log(SFR) density as a function of time (and redshift). It shows a dramatic rise in log(SFRD) from from ~ -1.8 locally ($z \leq 0.2$) to ~ -0.7 at

TABLE 1: Conversion Factors for Emission-Line SFRs

Line	$\mathcal{K}/10^{-42}$ erg sec^{-1}	Reference
H_α	0.79	Kennicutt et al. (1994)
O II	0.88	Hippelein et al. (2003)
O III	1.00	Hippelein et al. (2003)

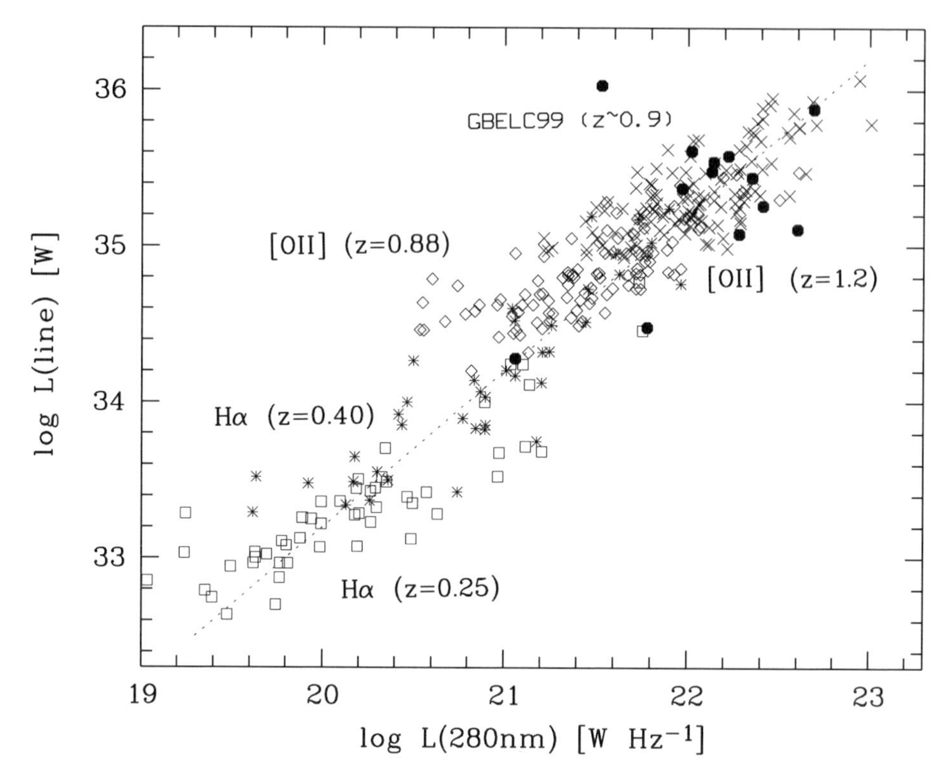

FIGURE 12: The UV continuum flux density at 2800 Å plotted versus reddening-corrected emission-line strengths for redshifts between $z = 0.24$ and 1.2, from Hippelein et al. (2003). The conversion to cgs units (erg s^{-1}Å^{-1}) is managed by multiplying by 2.61×10^{-5} when $\lambda = 2800$ Å, giving continuum fluxes $\sim 10^{17}$ erg s^{-1}Å^{-1} in the mid-range of the horizontal axis.

$z = 1.0$; an increase in excess of a factor of 10! It is in good agreement with the Hogg *et al.* paper.

There are several potential physical explanations – two favored ones have to do with the gas content as a function of time, and the extant merger frequency, both predicting a rapid decline of the SFR after the $z \sim 1$ epoch.

A possible area of concern – common to many SFR density determinations – is the extinction correction; a systematic error in the reddening correction with galaxy luminosity, for example, could invalidate some of the main result we have quoted above (i.e., the SFR density increase by a factor of 10 to 15 times).

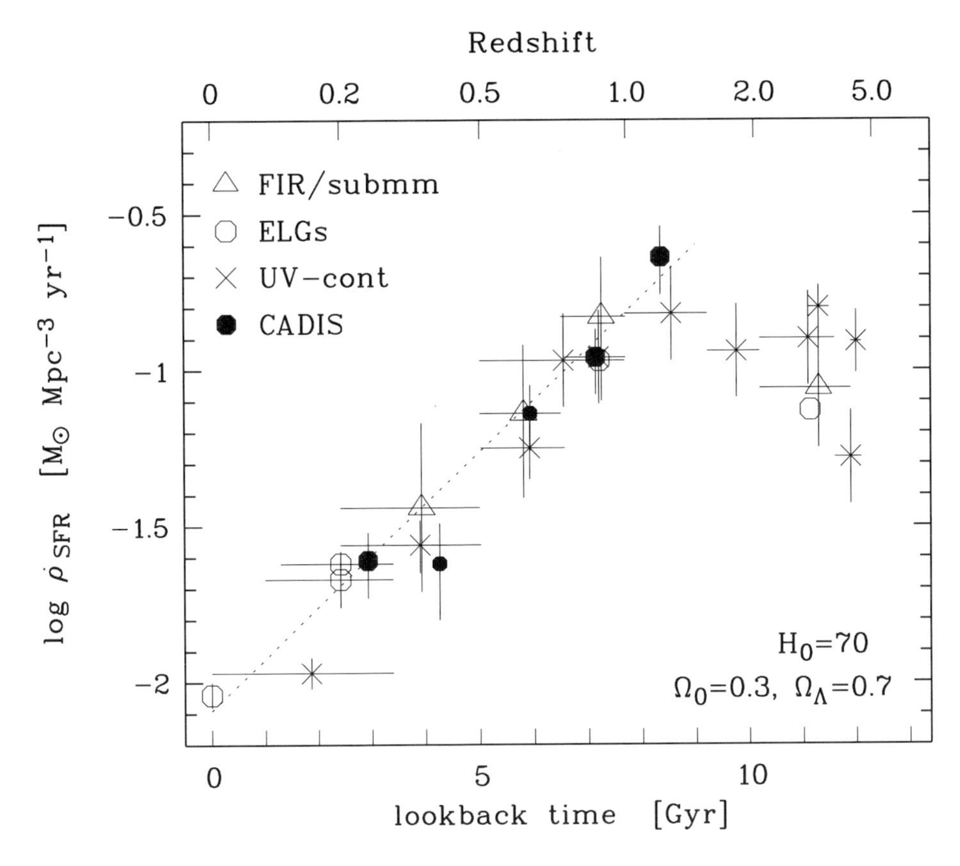

FIGURE 13: From Hippelein et al., Fig. 12; log SFR density as a function of look-back time for a flat Λ cosmology. The extinction-corrected CADIS data are represented by the filled circles, where the one at $z = 0.25$ is from Hα, the two at $z = 4$ and 0.64 are from [O III], and those at $z = 0.88$ and 1.2 are from [O II]. Extinction corrections increase with redshift. We refer the reader to Hippelein et al. (2003) for references to the sources of the other data.

3.10 Abundance Variations at Moderate Redshifts – Different Targets and Methods

The current crop of O and B stars in our Milky Way have an overall chemical composition that certainly suggests a global increase of metallicity with cosmic time. The oldest MW stars, globular clusters and possibly the field halo, are very deficient in heavy elements. However, detecting the systematic growth of stellar metallicity in the galactic disk is a much more subtle problem, even over a time-span in excess of some 10 Gyr.

It is difficult to map out the abundance behavior of differing extragalactic samples that would imitate the "time-line" of a typical spiral galaxy over the cosmic epochs.

Observationally we can select several classes of galaxies or "almost-galaxies" that yield reliable metal/H ratios (the nomenclature has designated [M/H] or [Fe/H] in logarithmic form, or even the linear ratio, $Z/Z_{\odot}$). We present these somewhat speculative correlations in Table 2. Some of these detections require a background AGN source of continuum radiation against which absorption can be detected. The approximate column densities of these different kind of systems (first column) correspond to the following types of physical systems:

- (a) Lyα clouds, low-density "fragments" (low-density columns; $\log(N) \lesssim 14.5$ (cm^{-2})
- (b) Lyα clouds $\log(N) \gtrsim 14.5$ (cm^{-2}); up to Lyman limit systems $\log(N) \approx$ 18 or so
- (c) Damped Lyα systems (substantial H-columns; $\log(N) \gtrsim 20$ (cm^{-2})
- (d) Lyman-break (SF) galaxies (stars and gas); normal galaxies
- (e) QSO and other AGN nuclei (central regions)

TABLE 2: Characteristic Large-scale density versus metal abundance level range

log Column (H I) (cm^{-2})	$Z/Z_{\odot}$	Baryon overdensity	Object class
$\lesssim 14$	$\lesssim 10^{-2}$	1–100	Lyα clouds
$\sim$ 14–18	$\sim 10^{-2}$	10^3	Larger Lyα clouds and Ly limit systems
$\sim$ 21	0.1:	$\geq 10^4$	Damped Lyα systems
21 to 22:	$\sim$0.5–1	$\sim 5 \times 10^4$	Normal SF galaxies
22 to 23::	1-10	$>10^5$:	QSOs nuclear regions

The rough metallicity in solar units is given in column 2; column 3 lists characteristic baryon over-densities, and finally, the object class is listed. Inspection of Table 2 shows an empirical trend between the over-density and the metallicity of our sample objects, as was also noted by Cen & Ostriker (1999)

The Lyα clouds with column densities $N \lesssim 3 \times 10^{14}$ cm^{-2} are generally thought to have metallicities $Z/Z_{\odot} \lesssim 10^{-2}$, though little is known of them in detail. For $N \gtrsim 3 \times 10^{14}$ cm^{-2}, metallicities are $Z/Z_{\odot} \approx 10^{-2}$ (Cowie et al., 1995).

In the two best-studied types of systems above, the normal interstellar gas of star-forming galaxies and the damped Lyα systems, we can (in principle) discern their behavior as a function of redshift. However, in the damped Lyα systems, one systematic uncertainty is simply our lack of robust (optical) identifications of the intervening absorbers, which appear to vary from dwarf, to low surface-brightness (LSB) galaxies, to possibly grand design spirals.

3.10.1 Spectroscopy of Star-forming Galaxies, and their Metallicities

If the damped component is part of an intervening "host", perhaps low surface-brightness galaxy, that would suggest a matching low SFR. Comparing the empirical evidence for the brightest Ly-break galaxies ($z \sim 3$, which puts the key absorption lines in a good spectral domain) to a nearby sample of starbursts, one can use the lines originating in the ISM of the "parent" galaxy. Heckman et al. (1998) observed a number of nearby actively SF galaxies (thus relatively bright in their UV), and identified spectral features arising from B stars, O stars (stellar wind lines), and the UV absorption lines due to the ISM in the hosts. Using the three fairly strong, low-ionization ISM lines just long-ward of Lyα ($\lambda\lambda$ 1260 of Si II, O I 1302, and C II 1335 Å) which appear to be moderately saturated, Heckman et al. (1998) formed a plausible correlation with the metallicities of nearby starburst galaxy spectra. However, the resolved high-ionization doublet Si IV $\lambda\lambda 1394, 1403$ strongly suggests line saturation, and that in turn implies a dependence on the detailed line width mechanisms and their magnitude, as much as on the element/H ratio. The mean EW of the three absorptions mentioned above ($\mathcal{W}_{\rm ISM}$ in the Heckman notation) varies from about 4.0 Å at the solar [O/H] abundance level to $\sim$ 2 Å for a metal-deficient system, down by a factor of $\sim$ 6 with respect to solar abundance. There are only a few $z \sim 3$ galaxies bright enough to yield adequately robust $\mathcal{W}_{\rm ISM}$ values on our Keck LRIS spectra; Table 3 lists the line strengths for these luminous galaxies – and we note several correlations in the data. These galaxies are called Lyman break galaxies (LBGs) because their continua characteristically show a break near and below the Lyman series, at rest 912 Å to 1216 Å, due to scattering of UV radiation by neutral hydrogen. We are aided in our need for bright sources with spectroscopy by the $z = 2.724$ Lyman break galaxy (see Chapter 4), ms1512-cB58, which is gravitationally magnified by a factor ≈ 30. (Pettini et al., 2002).

The mean EW of the three mentioned ISM absorptions correlated well with the oxygen abundance in the Heckman work, so we suspect that the five LBGs of Table 3 carry a substantial range in light element abundance, as their $\mathcal{W}_{\rm ISM}$ covers much of the range found by Heckman's team from slightly metal-poor in oxygen to "down" in O/H by perhaps a factor of 10–20 in the case of ssa-c17.

Research by Lilly et al. (2003) appears to show that distant normal spirals exhibit only slight abundance deficiencies for redshifts less than one. They examined the metallicities [O/H] in a large sample of star-forming galaxies over the intermediate distance range, $0.47 \leq z \leq 0.92$. The 66 galaxies, taken from the CFRS with measured optical emission lines, superficially appear to have spectra typical of spirals in our local Universe ($z < 0.1$). The Lilly team's main effort was devoted to utilizing the emission line fluxes to derive (slightly model-dependent) abundances, mainly the [O/H] ratio.

TABLE 3: Rest EWs of Absorption and Emission Lines in Five LBGs

LBG name	ms1512–cB58	hdf4–555.1	q0000 d6	hdfff d16	ssa–c17
Redshift	2.728	2.799	2.960	3.125	3.299
Damped Lyα abs?	yes	yes	no	no	no
Lyα abs	>8.18	—	0.95	1.62	—
ISM lines (Å)					
Si II 1260	2.88	3.20	2.17	2.15	1.52
O I 1303	4.56	4.90	2.52	2.02	0.89
C II 1336	3.55	2.42	2.19	1.35	0.40
SiII 1527	2.63	2.62	1.25	0.84	—
FeII 1608	1.30	1.32	0.52	0.56	—
AlII 1671	2.81	2.90	1.04	0.53	—
ISM 3 (ave.)	3.60	3.50	2.29	1.84	0.94:
Photospheric lines					
C III 1175	2.41	—	1.11	1.32	—
S V 1502	0.32	0.75	0.55	0.33	—
O–star wind line					
N V 1240 em	−0.62	—	−0.79	−1.93	—
SiII 1260 em	−0.16	—	−0.79	−0.40	—
O I 1302 em	~ 0	—	−0.86	−1.22	—
C II 1336 em	−0.10	—	0.25	−0.27	—
SiIV 1394 abs	2.16	1.42	1.66	1.17	—
C IV 1549 abs	> 4.98	2.92	5.86	3.35	1.5:
C IV 1549 em	−0.87	−0.66	−1.33	−0.71	—
Emission lines					
Lyα em	~ 0	—	−9.34	−11.75	−8.3
HeII 1640 em	—	−0.68	−0.87	−0.39	—
Continuum color (UV – qualitative)	red	red	sl. red	neutral	blue

* A minus sign for a line indicates emission.

The actual derivation of a metallicity level for oxygen utilizes Pagel's R_{23} estimator (Pagel et al., 1979),

$$R_{23} = \frac{f_{\mathrm{OII}\,3727} + f_{\mathrm{OIII}\,4959,\,5007}}{f_{H\beta}}, \tag{4}$$

where fs are line fluxes for visual-region emission lines. Metallicity solutions were calibrated by McGaugh (1991), as shown in Fig. 14

Lilly et al. (2003) use this relation to measure the differential trend of [O/H] with redshift; the value for a given galaxy is affected by the degeneracy in the R_{23} metallicity solutions at an "abundance turnaround" at [O/H] = 8.4, where $[\mathrm{O/H}] \equiv \log(n(\mathrm{O})/n(\mathrm{H}))+12$. [O/H] as a function of R_{23} (see Fig. 14, is calibrated by McGaugh 1991). We note parenthetically that the abundance ratio between oxygen and hydrogen in our Sun still seems to vary somewhat by technique; now it seems preferable to describe the logarithmic solar abundance by [O/H] = 8.7;

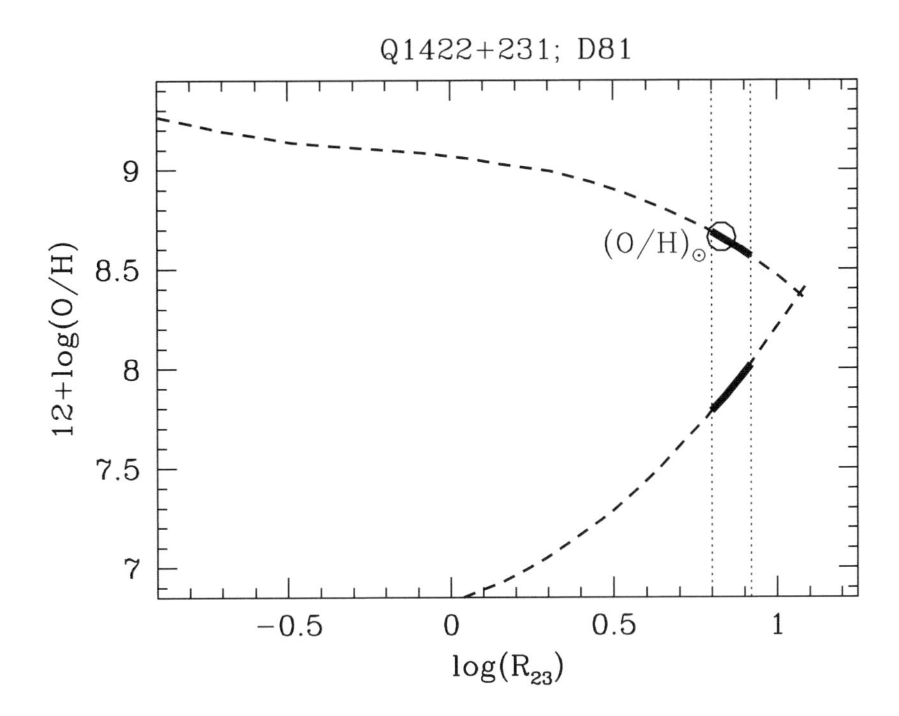

FIGURE 14: The oxygen abundance solutions, as calibrated by McGaugh (1991), are shown as a function of R_{23}. The vertical dotted lines are the range of $\log R_{23}$ values consistent with the from the galaxy Q1422+231 at $z = 3.1037$. The open circle represents a recent determination of the solar oxygen abundance, $12 + \log(\mathrm{O/H}) = 8.66$ (e.g., Allende-Prieto et al., 2001). The solution for a given R_{23} is 2-valued; in this case, the difference in (O/H) solutions is a factor of ~ 5. Note that the R_{23} method is simple, but not unequivocal.

the Lilly *et al.* (and other data sets) often used 8.9. To lessen the above [O/H] degeneracy, Lilly *et al.* attempt to add measures of the two red lines, Hα and [N II] 6584. Owing to the substantial strength of the [N II] line, the derived ratio [N/H] implies that the metal-rich interpretation from the bimodal oxygen R_{23} solution is probably the correct choice; that is, [O/H]$>$ 8.4. The nitrogen lines were only observed in six galaxies, however, so the diagnostic assumption was made in $\sim$10% of the galaxy sample. The cumulative data are all consistent with [O/H]$>$ 8.4 – in the upper portion of Fig. 14.

Following the assumptions of the Lilly group, this analysis implies only a modest change in metallicity from $z \sim 1$ to $z \sim 0$; Δ[O/H]$= 0.08 \pm 0.06$. Thus, little change in galaxy metallicity accompanies star formation over the last half of the cosmic age more recent than unit z.

In contradistinction to the conclusion that the metal-rich solution is most often correct, occasional small star-forming systems that are quite nitrogen-poor galaxies have been found (Martin & Sawicki, 2004; Stockton & Ridgway, 1998).

In summary, with some room for re-evaluation, we find a more or less constant metallicity for SF galaxies out to $z = 1$, but with some rare exceptions. However, at redshift $z \sim 3$, most of the LBGs are probably moderately metal-poor.

3.10.2 AGN Chemical Abundance Evolution

The presumed high-density environment of galaxy nuclei in the material currently being accreted into the proximity of a MBH, perhaps especially for AGN, occupies a small volume in the Universe, and may often be super-solar in "light element" abundances.

QSOs are bright and small, so studies of their metallicities and metallicity gradients can, in principle, provide measures of the gas-phase abundances and densities. These densities are likely large if the creation of a large BH also involves extensive SF in the central regions of the host galaxy.

So we are interested in how large the typical the abundance ratios (O/H), and (N/H) become in QSOs over cosmic time, and whether a dependence on redshift can be recognized. The latter question may be difficult to answer, as the spectral analysis may depend upon the (redshift) placement of key ions in the available observational windows.

In particular, the [N/H] ratio in QSOs is likely to reflect the astrophysically secondary enrichment of nitrogen in the stars and the nuclear gas: the rest-frame spectral features of high priority being N III]1750 Å, and N V 1240 Å. These lines generally represent gas in the broad emission-line region (BELR) of the QSO. For contemporary ground-based measures of the above N-ions we select $z > 2$ QSOs to allow measurement of the redshifted N lines well-above the atmospheric cutoff of Earth's opaque O_3 bands.

A common interpretational procedure emphasizes also measuring the appropriate oxygen ions, since in nearby H II regions the N/O abundance scales with the O/H abundance ratio. This behavior occurs in "secondary" element production; N is synthesized in stars of mass $M \gtrsim 5\ M_\odot$ from pre-existing C and O nuclei produced in earlier CNO-burning. As illustrated in Hamann & Ferland (1999), the relative nitrogen abundance (N/O) can be a useful guide to the overall heavier metallicity like (Fe/H), which is more difficult to measure in distant QSOs. This suggests using oxygen ion lines at $\lambda_0 = 1664$ Å and 1034 Å, which again are most commonly available at high redshifts, say $z \gtrsim 3.0$.

If we could measure a redshift gradient in the abundances of N and/or O, we would likely learn of the SF lifetimes required to build up the likely reservoir of metal-rich gas we actually observe. Failing this, simply observing young, high-z QSOs may place limits on the nucleosynthetic stellar timescales. That method has become popular since large telescopes can better observe QSO spectra to the faint levels seen for $z \geq 4$ QSOs; some limited HST data is available for local QSOs – see Laor et al. (1994).

The quasar elemental abundances at high z have recently been examined by Dietrich et al. (2002), and Dietrich et al. (2003). With especial attention paid to the N V 1240 ion line in the BELR, these teams of investigators confirmed that

Chapter 1

FIGURE 1: The Milky Way, seen in the 2-3 μm near-IR with the COBE satellite. Image courtesy of Mike Hauser and the COBE collaboration. See text for discussion.

FIGURE 3: The Large Magellanic Cloud, an irregular galaxy beginning to merge with the Milky Way galaxy. This "almost Irr" galaxy is close enough as to be a valuable yard stick in the search for a consistent extragalactic distance scale. Image taken with the Anglo-Australian Observatory's Schmidt camera; courtesy of Anglo-Australian Observatory/David Malin Images.

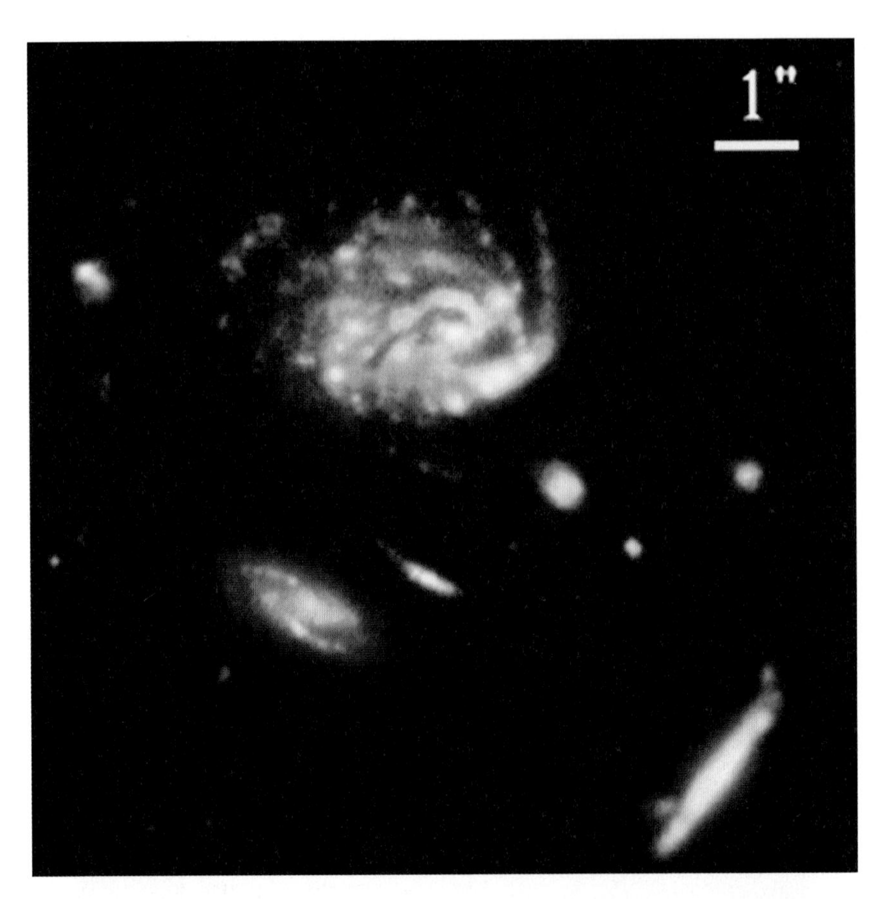

FIGURE 6: An enlargement of a portion of the Hubble UDF (a product of the HST and the Space Telescope Science Institute; STScI) showing a coalescing group of galaxies at $z \approx 1$; image courtesy of H. Yan and R. Windhorst. The face-on spiral seems perturbed and asymmetrical. The edge-on spiral appears to be in the process of cannibalizing two dwarf galaxies. The red galaxy appears to have the morphology of a spiral; the highlight on its lower arm suggests a burst of SF. Other smaller galaxies in this image have sizes of about $\sim 0.2''$, suggesting a physical size of about 1.6 kpc at $z = 1$. One must remember that most mergers are of unequal sized objects.

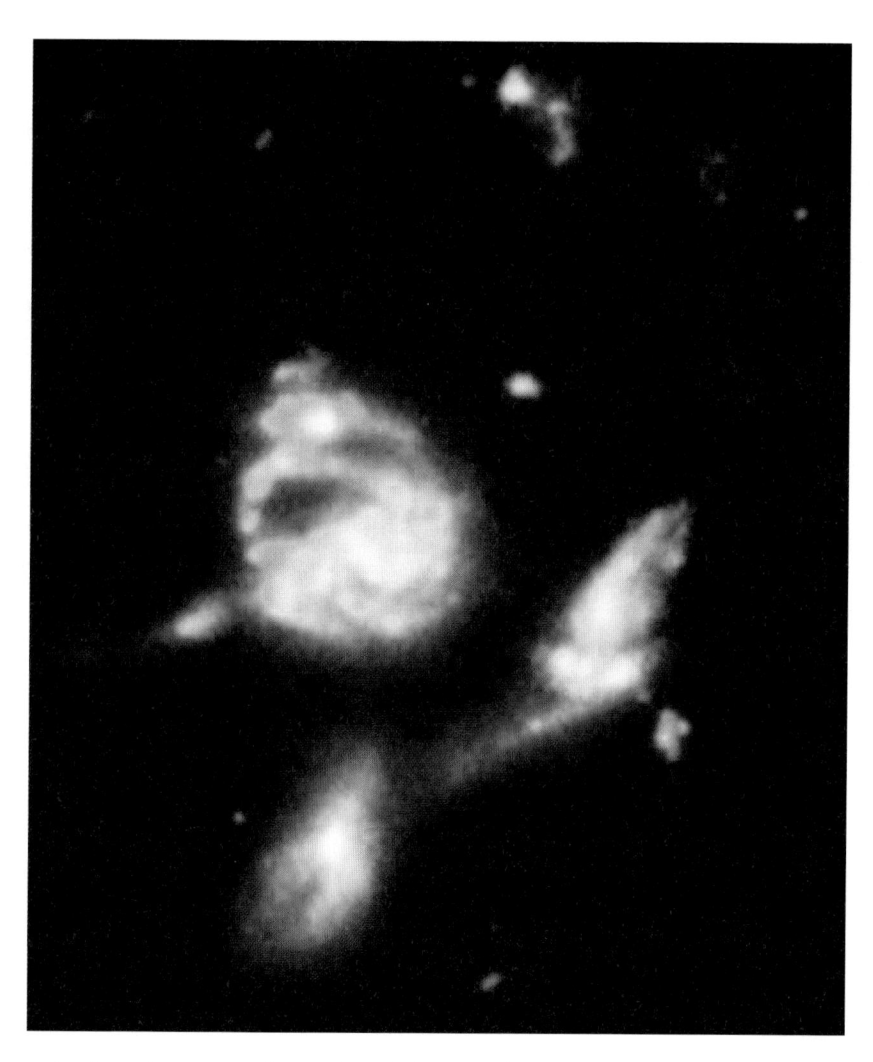

FIGURE 7: An enlargement of a portion of the UDF (courtesy of H. Yan and R. Windhorst) showing a coalescing group of galaxies at $z \approx 1$. Though it is difficult to be sure, each of the major galaxies appears disturbed, probably from interactions. Notice the large number of very small galaxies around this group – dwarf galaxies which are probably physically associated with the coalescing group.

Chapter 2

FIGURE 1: NGC 4676A,B, taken with the HST Advanced Camera for Surveys (ACS) showing the results of a recent interaction of two spiral galaxies. These galaxies are thought to be on the edge of the Coma cluster. Photo from NASA ACS Science and Engineering Team, NASA and the Space Telescope Science Institute (STScI).

FIGURE 2: The polar-ring galaxy NGC 4650A; a Hubble Heritage image. These systems are thought to be formed by galaxy interactions; perhaps a "direct hit", in the above figure. The relatively large frequency of polar-ring galaxies suggests that they are dynamically relatively stable, compared to their non-polar cousins. Image courtesy of STScI.

Chapter 3

FIGURE 9: The nearby spiral NGC 7424, taken with VIMOS/VLT at the European Southern Observatory. Note the prominent bar; the redder colors of the bar (see color section) indicates an older population of stars populate the bar. Such features are difficult to detect at high redshift because our optical image then views the rest UV, and old stars emit very little UV. Imaging in IR bands may be of use in detecting bars at high redshift, as per the work of Bunker et al. (2000).

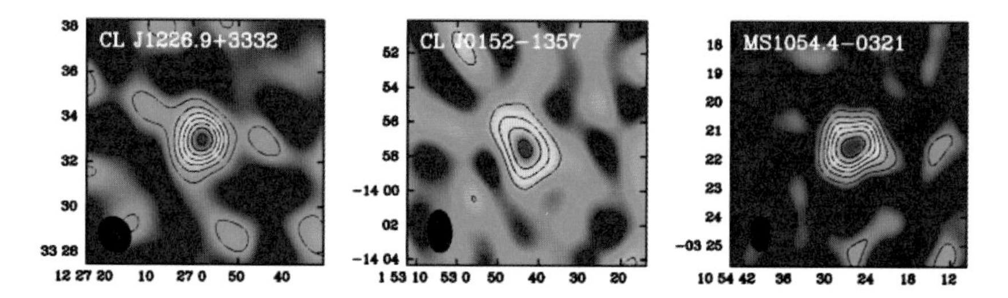

FIGURE 10: Synthesized images of the SZ effect for three cluster fields. Contours represent multiples of 1.5-σ, where σ is the RMS flux. The data can be used to fit a cluster electron density model to the SZE decrement (usually an isothermal β model) to help derive total masses and cluster temperatures of the clusters. Derived masses for each of these clusters are $\gtrsim 2 \times 10^{14}$ M$_\odot$within 340 h$_{100}^{-1}$ M$_\odot$. Figure from Joy et al. (2001).

Chapter 5

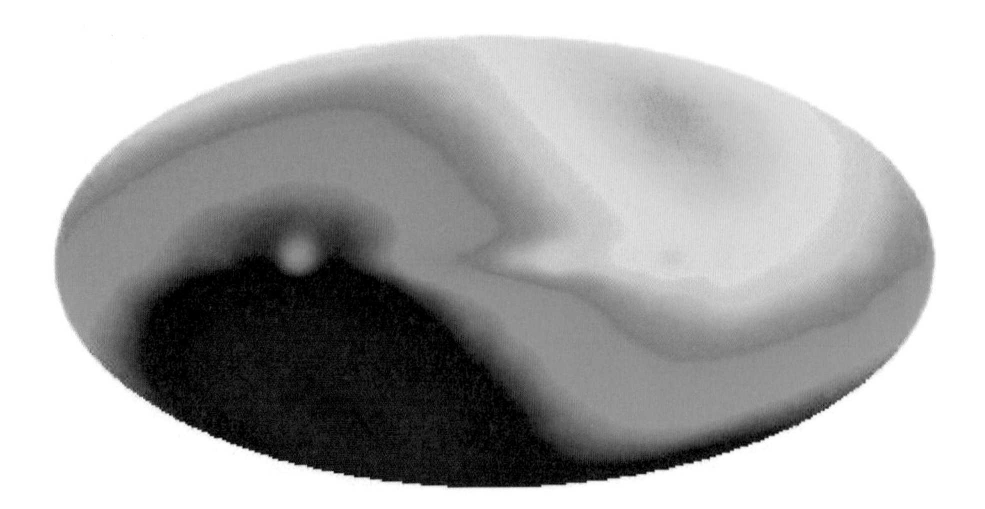

FIGURE 3: The dipole anisotropy, detected with the differential microwave radiometer on COBE (Smoot et al., 1991). The Galactic coordinates $(l, b) = (0, 0)$, is located at the center, and b increased from zero toward the left. The CMB temperature is higher in the upper-right, indicating owing to our motion in that direction (galaxy rotation, local group motion) relative to the cosmic rest frame.

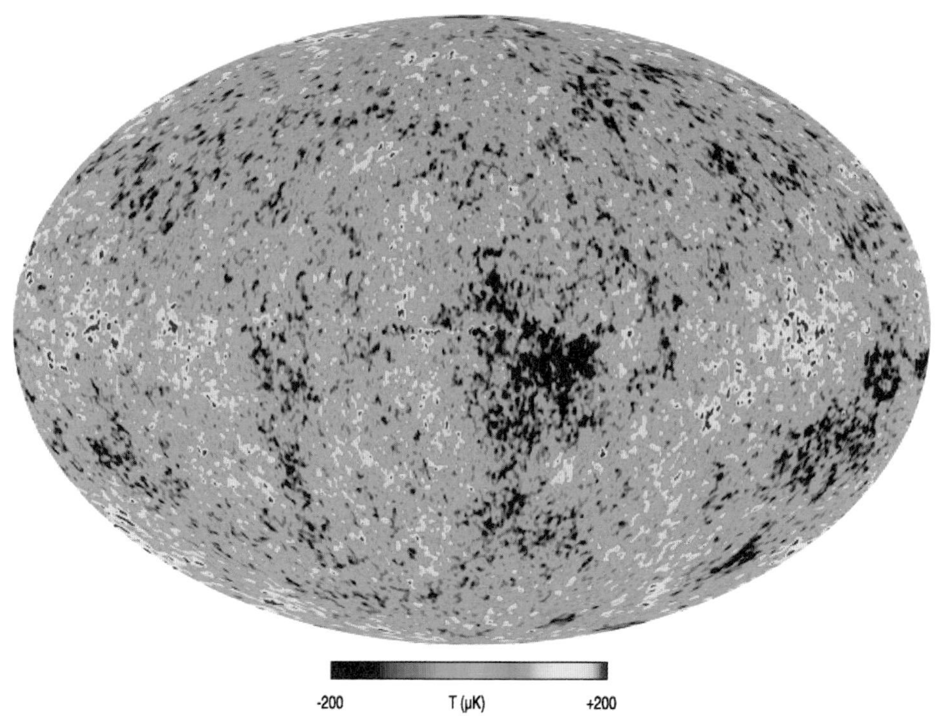

FIGURE 4: "Internal linear combination" map combining the five band maps of the WMAP data in such a way as to subtract "foreground" contaminating fields, such as various features of our Milky Way galaxy, while constraining the mean CMB flux to be constant. For a more detailed description, see Bennett et al. (2003). The plane of the galaxy is horizontal, the center is at a longitude of 0 degrees.

Chapter 6

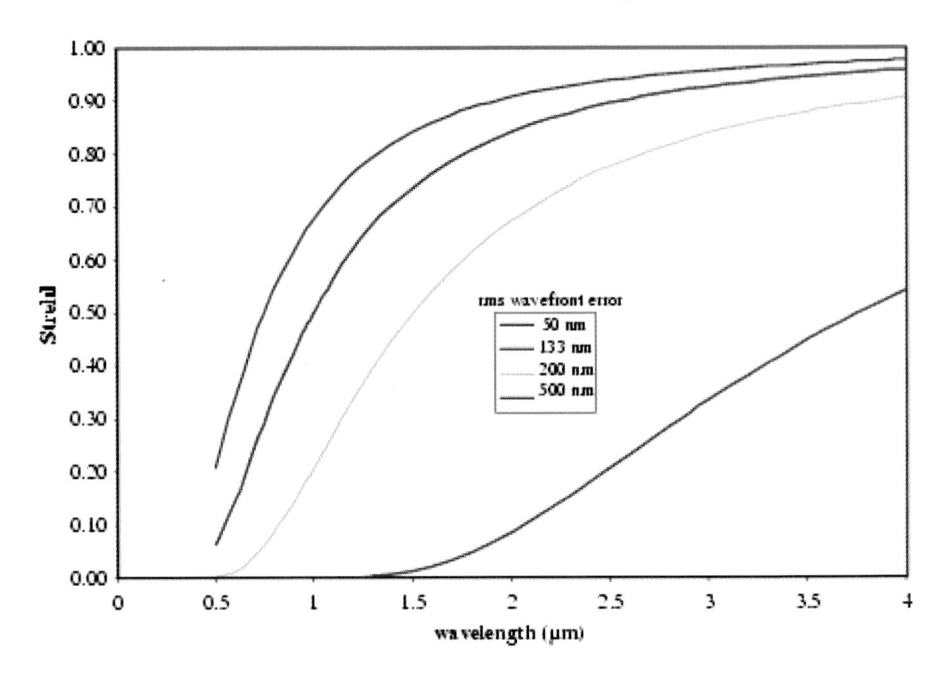

FIGURE 2: Strehl ratios as a function of wavelength of light, shown for various wavefront errors, given in micrometers. Graph from Nelson & Mast (2004). Smaller wavefront errors result in larger Strehl ratios.

at moderate and even high redshifts ($z > 4$) the N/H ratio in the QSOs ranged around supersolar levels. The mean QSO metallicity was $Z/Z_{\odot} = 4.3 \pm 0.3$, at $z \geq 4$. This represents a rapid SF and metal-production epoch early in the life of a (bulge) population of stars. A rare case (Q0353-383) studied by Bentz & Osmer (2004) and Baldwin et al. (2003) suggests an N overabundance of 15 times solar! We can utilize the abundance excesses to estimate an evolutionary timescale for the chemical enrichment of the dense nuclear BLR gas. Dietrich et al. (2003) study the Fe II/Mg II ratio among QSOs from $z \sim 0$ to 5 and find a lack of evolution in the ratio. The chance for detecting evolution appears hinged on the different timescales for enrichment of alpha elements like Mg, which are formed on short timescales in massive type II SNe, and iron, formed in type II SNe. Thus, the iron enrichment is expected to have a delay of 0.2 to 0.6 Gyr. Applying this timescale to an observed QSO redshift of $z \simeq 5$ we find that massive SF (to also produce heavy elements) commenced at a redshift of $z_{\rm form} \simeq 7 \pm 1$. That epoch is also consistent with the end of cosmological re-ionization. A desideratum for the future will be the measurement of nitrogen in BELRs at $z > 6$, for to build up a MBH toward 10^9 $M_{\odot}$requires a rapid and early "feeding" stage.

A new angle that may concern us is the physical extent of the super-solar high-abundance region: a recent suggestion by Kuraszkiewicz & Green (2002) that C, N ion narrow absorption lines are found in larger-scale paths in the host galaxy, not just involved with the 0.1 pc scale BELR. Of course we do not know the 'ancestry' of this absorbing narrow-line gas. It could have its origin in the Super-Metal-Rich spheroid stars or perhaps was (an earlier) nuclear ejection and thus would be foreign to the larger-scale metallicity suggestion.

Hopefully this potentially valuable "side-show" dealing with the physical scale of the metal-rich gas in AGN may be explored more-fully in the future.

3.10.3 The Abundance Evolution of Damped Lyα Absorption Systems

The so-called Damped Lyα (DLA) spectral signature is connected to a strong H absorption at the Lyα line itself. It indicates large H I columns; $N(\mathrm{H\ I}) \geq 2 \times 10^{20}$ cm^{-2}. Along with the strong H I absorption (with damping wings) we also detect UV absorption lines due to once-ionized metals like Fe II, Zn II, Cr II, Al II, and Mg II. Thus, in the case of the damped absorbers, our sight-line to a background QSO source traverses a substantial amount of enriched gas, whose origin and physical state is still uncertain. But we can measure with difficulty the abundance ratios for several ions in most of these cases.

In this chapter we are concerned about the gas phase metal-to-H abundances of the species that are non-refractive. The depletion onto grains for refractory elements like Fe weakens our usual habit of equating [Fe/H] to the overall metallicity of the environment; iron is fairly abundant in our MW Galaxy, but well-locked to ISM grains in the galactic disk. Zn is better; it is non-refractory and stays in the gas phase. Thus for the simplest gas-phase abundances, with minimal (or zero) correction for dust adhesion, the [Zn/H] ratio will be the ratio of choice.

However, it is difficult to acquire (Zn/H) data over the redshift range $z = 0$ to $z = 1$, as the Zn II lines are located near $\lambda_0 \sim 2000$ Å, well into the spacecraft UV for low-z candidate objects. Thus, few QSOs with damped Lyα and a coincident Zn II metal system are found below $z = 1.0$. A very few are available.

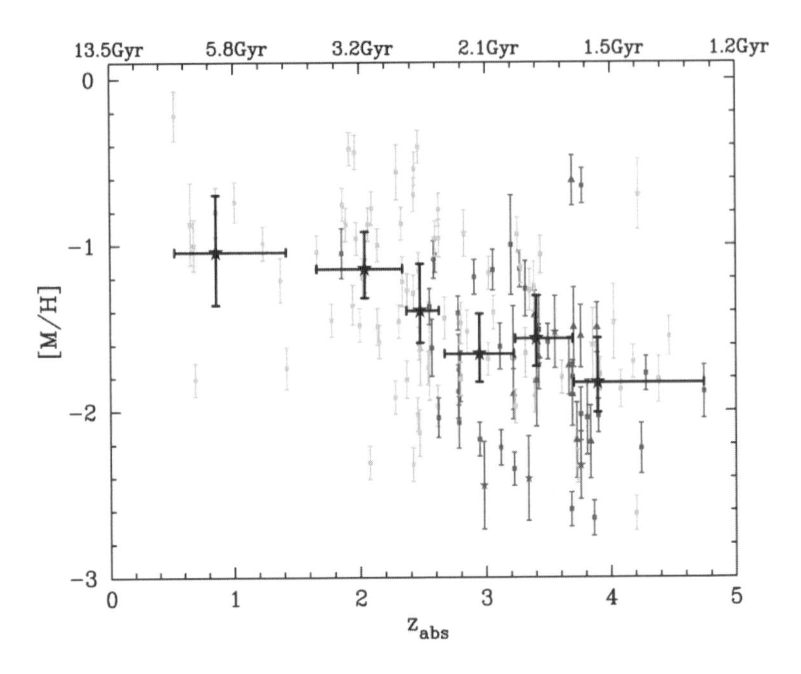

FIGURE 3: Metallicity, in logarithmic form, vs. redshift for 125 DLAs of the sample of Prochaska et al. (2003). The unbinned (in z) points are individual measurements, and the bold-faced points represent the binned cosmic mean metallicity $\langle Z \rangle$, representing the metallicity of the universe in neutral gas in over-dense regions. While the lower axis shows redshift, the upper horizontal axis shows the age of the universe for a flat Λ cosmology.

Figure 3, from Prochaska et al. (2003), shows the empirical evidence on the overall gas phase metal abundances (with Fe corrected for dust depletion) for 125 damped DLAs, where "stars" with error-bars represent redshift-binned mean values. When possible (most of the time), species that are less refractory than iron, such as Si, S, O, or Zn, were used to estimate the metallicity. Only a few DLAs lie "inside" of $z = 1$ and thus to state a realistic change in abundance over late cosmic epochs ($0 < z < 1.0$) requires an unwarranted extrapolation to the $z = 0$ conditions (and Zn/H) observed in the young stars of our own MW galactic disk. There is only a slight decrease apparent in [M/H] from $z = 0.5$ to $z = 1.0$.

However, we note that our interpretation of the DLA gas clouds is physically clouded by our lack of robust identifications for these sight-line absorbers. The DLAs could be small galaxies, pre-galactic disks, or LSB galaxies (Vladilo, 1999). Vladilo et al. (2000) and Edmunds (1999) discuss the possibilities; we can often find one or more attributes of normal spiral galaxies in the observed gas DLA-properties, but not necessarily a majority of the same. The DLA's characteristics include:

- a fairly large covering factor for optically thick H I,
- a likely *low* SFR, since we rarely see substantial optical/UV radiation associated with the absorber position and redshift,
- low ionization states for the metals; they are ionized by a galaxy-like (soft) radiation field,
- an absorption-line velocity dispersion suggestive of a galactic disk,
- and an overall metallicity somewhat below solar; dust may be present.

But no one specific and unique galaxy fits all of these five characteristics, and is identified with a DLA. The subject is an active one, so improvement may well occur in the near future. Optical identifications of $z < 1$ DLAs suggest a significant fraction of DLAs are small or LSB galaxies and, perhaps less likely, spiral galaxies (Vladilo, 1999). If DLAs are composed of a substantial population of LSB, or pre-galactic stages, then perhaps we should not be surprised if metallicities do not evolve very rapidly.

Very recently, Möller et al. (2004) succeeded in detecting Lyα emission from the center of some high-z DLAs. Their study finds that out of 17 DLAs studied, 5 have localized Lyα in emission within $1''$ of the QSO line of sight. All of the five DLAs appear to have metallicities, determined by the QSO spectrum, larger than the mean of the 17. They conclude that Lyα emission appears to correlate with metallicity in the sense that the SFR correlates with metallicity and Lyα emission correlates with SFR. They suggest that most DLAs are smaller than typical LBGs, and tend to be obscured, in conformity with our tentative conclusions above.

3.11 QSOs, AGN: Evolution at Large Redshifts

The evolution of AGN with redshift, even when restricted to cover the moderate z-range $0.2 \leq z \leq 1.2$, is very striking; perhaps it is indicative of the largest decline in activity with cosmic epoch we have explored in this chapter. Certainly our straightforward modeling suggests that the AGN/QSO luminosity function, spelled out most thoroughly through the QSOs, evolves rapidly with cosmic epoch. The present epoch (here and now) is relatively quiet; luminous AGN being very rare locally.

One fairly simple test, invented by Schmidt (1968) is called the luminosity–volume test, or in the modern literature, the mean $V/V_{\rm max}$ statistical ratio test. For each source in our sample we form a (cosmologically consistent) volume at which it's distance is measured from a redshift and the volume to which it would barely be observed (deriving an $r_{\rm max}$ distance). Then that yields a $V_{\rm max}$ (waveband selected). If we consider an ensemble of physically similar objects, and if we expect their distribution with redshift to be uniform, then half of the sources would be expected to be found in the inner half of the maximum volume, and the rest in the outer half. Thus the ideal resultant statistic, $\langle V/V_{\rm max}\rangle = 0.5$, where the angle-brackets indicate an ensemble average. We note that multiple selection criteria – observations in more than one wave-band – can be used on thoroughly studied distant objects, such as QSRs (radio-loud, and optically bright) for multiple $\langle V/V_{\rm max}\rangle$ distributions. Some $V/V_{\rm max}$ results (mainly from Peterson 1997) for QSRs and QSOs are listed in Table 4.

TABLE 4: Selected $\langle V/V_{\rm max}\rangle$ Results for AGN Samples

Sample	$\langle V/V_{\rm max}\rangle$	N	Reference
3CR QSRs	0.694 ± 0.042	34	Wills & Lynds (1978)
4C QSRs	0.698 ± 0.024	76	Wills & Lynds (1978)
X-ray QSOs	0.660 ± 0.062	22	Weedman (1986)
Grism Survey, QSOs, C I, $z = 2$ to 3.2	0.528 ± 0.036	72	Schmidt et al. (1991)
Grism Survey, QSOs, Lyα, $z \geq 2.7$	0.377 ± 0.028	90	Schmidt et al. (1991)

We note that implicitly the first three entries, which are dominated by QSOs/QSRs at redshifts less than 2, yield $\langle V/V_{\rm max}\rangle$ values far above the uniform distribution target of 0.50. The last entries suggest a substantial fall-off in the QSO space density at $z > 2.5$. These early trends have been confirmed by several contemporary studies.

As QSO samples have grown, our ability to observe and parameterize their luminosity function in several coarse redshift bins has also matured. In Fig. 4 one can see their consistent and fairly dramatic evolutionary signature. The plot presents comoving number densities of QSOs as a function of redshift-binned absolute magnitudes, for an Einstein–de Sitter Universe (Boyle et al., 2000). The QSOs come from the 2dF galaxy redshift survey, and contains $\sim$ 6000 QSOs. Note, for example, that the lowest redshift bin effectively has a negligible population of luminous ($M_{\rm B} < -23.5$ mag) candidates, while at $z \sim 1.3$ the QSOs are observed to much higher luminosities, to $M_{\rm B} \sim -26$.

Early work showed a "flux-deficit" on the blue-side of the Lyα line of high-z quasars (Oke & Korycansky, 1982; Schneider et al., 1989) resulting from scattering of Lyα photons by neutral intergalactic hydrogen clouds. We encounter this again in Chapter 4, where very high redshift quasars are observed to have a complete flux deficit on the blue-side of the Lyα line.

The essentials of the space density evolution of luminous quasars as a function of redshift was also an early finding. It was shown (Schmidt et al., 1988; Warren et al., 1988) that comoving number densities of bright quasars increase going from $z = 0$ to $z \sim 1.5$, or so. There is also a strong decrease in their comoving densities

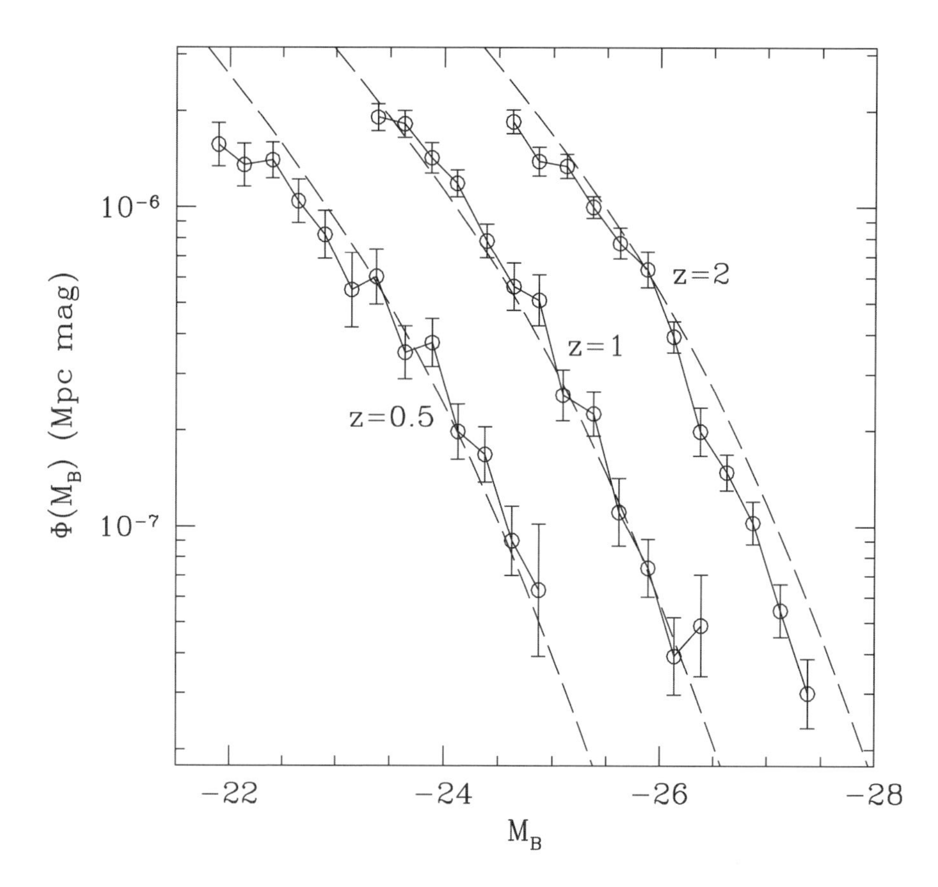

FIGURE 4: The luminosity function for the 2QZ+LBQS data set in a flat $q_0 = 0.05$ universe. Data for $z \approx 0.5$, 1.0 and 2.0 is shown in relation to the modeled QSO luminosity functions (Boyle et al., 2000).

for $z \gtrsim 2.5$ or 3. We shall see that this has major ramifications for the re-ionization of the Universe.

The luminosity functions from the more modern 2dF QSO Redshift Survey (2QZ; Folkes et al. 1999) can be parameterized by two power-laws, with slopes $\alpha = -1.58$ and $\beta = -3.41$, so that a satisfactory fit is achieved over a large range in redshift (Boyle et al., 2000). Figure 4 shows data gathered from the 2QZ for three redshift bins at $z \approx 0.5$, 1.0 and 2.0, with corresponding plots from the model. Owing to the size of the 2QZ, luminosity functions can be calculated in bins from $z = 0.4$ to $z \gtrsim 2$ (the initial photometric selections prevent this program from studying QSOs at $z > 2.5$).

If we simply examine the histogram of the number of QSOs surveyed (with mild corrections for incompleteness) with redshift (Boyle et al., 2000), we note the apparent extreme paucity of nearby QSOs. At $\overline{z} = 0.5$ the 2dF yielded $\lesssim 20$ QSOs; at $\overline{z} = 1.1$, that number grew to ~ 160 QSOs, about a factor 8–10 increase over the number for $z < 0.5$. However, if we correct for the comoving volume per redshift bin (see Fig. 2), the local deficit drops to a factor of ~ 3. This is still significant.

At the bright end, the difference between low and high redshift is somewhat more dramatic; at $M_{\mathrm{B}} = -24$ they are $\sim 2 \times 10^{-7}$ QSO/Mpc^3/mag at $\overline{z} = 0.5$. At $\overline{z} \approx 1$, there are 10^{-6} QSO/Mpc^3/mag in the same absolute magnitude bin; a factor of $\sim$5 larger, showing the late time decline in the LF at the bright-end.

The best interpretation and explanation of this evolution of data on AGN is evolution – either density or luminosity evolution. With density evolution, the comoving number density of QSOs changes with redshift. It would be represented in Fig. 4 by an up/down shift. The absence of very high-luminosity QSOs near us would then be due in part to their intrinsic rarity, and the limited volume available at low-redshift ranges, and perhaps luminosity evolution.

In the case of luminosity evolution we envision a horizontal shift, with the number of AGN being constant over cosmic time, but with the luminosity decreasing as we approach the present epoch ($z \ll 1$).

On the bright end of the luminosity function distributions we note a better fit with pure luminosity evolution. The best-fitting models suggest an exponential increase in luminosity as a function of look-back time. It may be that density evolution is plausible at the "faint" end ($M_{\mathrm{B}} > -23$). Boyle considers pure luminosity evolution models to be responsible for the observed evolution of the quasar luminosity function.

A recent, more complete, extension of these results was made by Croom et al. (2004), who use 23,338 QSOs from the 2QZ survey, nearly four times the number available to Boyle et al. (2000). Derived parameters for the QSO distribution function in time have values quite comparable to Boyle et al, even though there has been a welcome shift from an Einstein–de Sitter, to the now more acceptable ΛCDM cosmological model.

Some other AGN display a rapidly evolving luminosity function, perhaps like the radio-loud quasars. Among the extragalactic radio sources, the powerful radio galaxies exhibit similar order in their optical/radio luminosity functions. The time or redshift change in the radio luminosity function can be described by pure luminosity evolution (Dunlop, 1999). Longair (1998) notes that the comoving luminosity density of steep radio sources attains a maximum at $2 \lesssim z \lesssim 3$, and decreases rapidly toward larger redshifts. The characteristic luminosity for QSOs, $L^*(z)/L^*(0)$, increases dramatically with redshift, while the redshift at which the L_{B} for quasars is maximum is at $z \simeq 3$ (Boyle et al., 2000), comparable with that of radio sources.

Haehnelt et al. (1998) use simulations and the Press–Schechter formalism, a methodology for calculating the mass-function of collapsing structure in the early universe (Press & Schechter, 1974), to model the birth of quasars. They find a rapid increase in the comoving density of quasars in time followed by a rapid fall. Haiman & Hui (2001) suggest that the duty cycle may be a function of $M_{\rm bh}/M_{\rm halo}$, so that lower values to this ratio have correspondingly longer duty cycles, and sub-Eddington luminosities (see Eqs. (9) and (10)). This would account for the drop in the numbers of bright quasars at low z, and allow for a growing population of low-luminosity AGN.

Because of the penetrating power of X-rays, the XLF is not as much affected by circumnuclear dust as the optical LF. Barger et al. (2003) point out that while powerful X-ray AGN ($L_{\rm X} \gtrsim 10^{45}$ erg s^{-1}) attain a maximum number density within the range $1.5 \lesssim z \lesssim 3$, AGN in the range $10^{43} \lesssim L_{\rm X} \lesssim 10^{44}$ erg s^{-1} have comoving number densities that increase gradually from large redshifts to the present. When AGN are divided into broad-line (quasars) and narrow-line (obscured) AGN, the comoving number density of narrow-line (obscured or Type II) AGN increases with decreasing redshift, but the broad-line AGN hold remarkably constant (Steffen et al., 2003), as would a population with no density evolution. However, the mean luminosity of the broad-line AGN is found to decrease with decreasing redshift, reproducing the strong apparent evolutionary trend for absolute magnitude-limited surveys of quasars. Narrow-line AGN increasingly dominate the XLF at low redshift (Steffen et al., 2003).

Under hierarchical structure formation, the largest objects formed late, and smaller things formed early. But we have seen that the low-luminosity LF from smaller halos is undergoing positive density evolution. What could be the cause of this "anti-hierarchical" behavior? Granato et al. (2001) suggest that the flow of matter to the AGN is regulated by the AGN luminosity to an extent dependent on the potential well of the galaxy halo. In a scenario laid out by Cattaneo & Bernardi (2003), AGN in giant ellipticals are fueled at the Eddington limit, but for AGN in weaker potential wells, such as spiral galaxies, accretion rates are substantially lower, $\sim$0.1 $L_{\rm Edd}$. The Magorrian et al. (1998) relation has shown that the BH mass bears a remarkably constant ratio to the mass of the host galaxy spheroidal component (bulge), at about 0.6%. This is also known as the $M_\bullet$-σ relation (Shields et al., 2003), where σ is the velocity dispersion of the spheroidal component and $M_\bullet$ is the BH mass. Shields et al. (2003) find that the local $M_\bullet$-σ relation holds constant up to $z \sim 3$. If true, the restraint on the growth of the BH brought about by the AGN luminosity and attendant feedback of mass flows, must produce a corresponding and proportionate restraint on SF in the bulges of spirals. The suppression of BH growth by feedback in narrow-line AGN results in a continual growth of the obscured AGN population over the whole range of redshift $z \lesssim 6$. These AGN can be directly detected with Chandra imaging of hard X-rays. The simulations of Cattaneo & Bernardi (2003) support this anti-hierarchical evolution of BH masses.

These results show that optical quasars are biased tracers of the SMBH accretion rate, for though the optical LF peaks at $z \sim 2.5$, the the lower-mass narrow-line AGN dominate accretion at $z \lesssim 1.5$. Since only $\sim$20% of the point-source X-ray background is accounted for by optical quasars (Salucci et al., 2000), most X-ray sources must come from obscured AGN. These general conclusions are in concert with those of Steffen et al. (2003)l; Type II (obscured) AGN densities increase with time, while broad-line AGN peak, then decline in their mean luminosity below redshifts of $z \sim 2$.

Thus, it appears that obscuration strongly affects the number counts of low-luminosity optical AGN. Working with the Chandra observations of the Hubble Deep Field North, Cowie et al. (2003) showed that bright X-ray luminous objects are more frequently observed as broad-line AGN, while less luminous objects are more often obscured, showing absorption features in soft X-rays. This suggests that feedback, the self-limiting effect of energy release, is more effective in limiting the accretion rate of the BH when potential wells are less-deep.

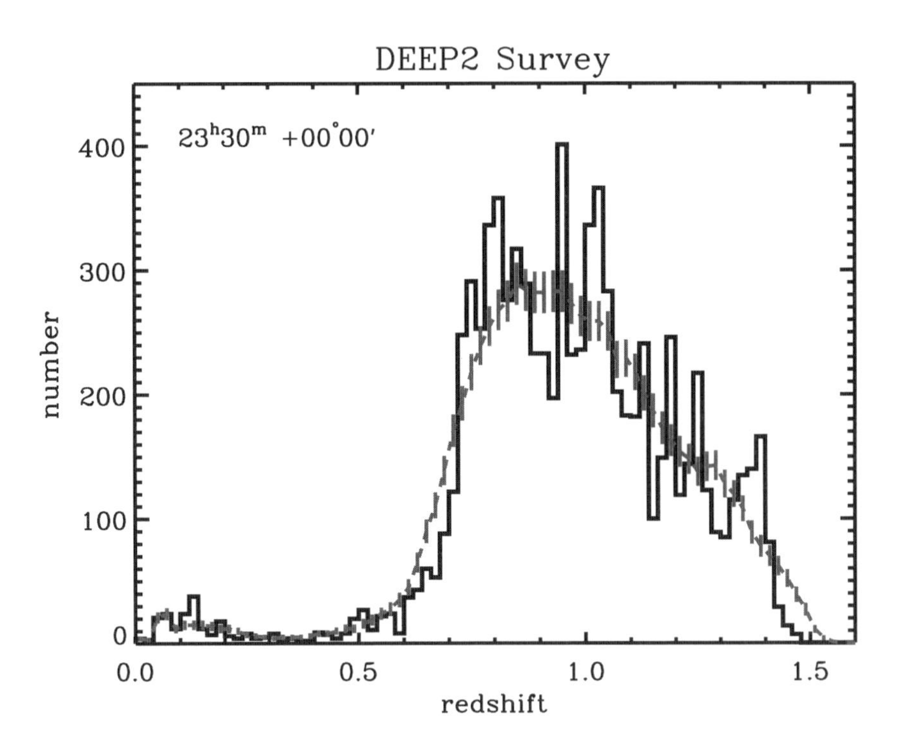

FIGURE 5: The histogram of galaxy redshifts from one of the early DEEP2 Survey fields. To display the significance of the apparent clustering, the galaxy distribution was smoothed, and the 1-σ Poisson errors were calculated, and overplotted, centered on the mean distribution. The overall shape of the redshift function is largely determined by the selection criteria, mainly apparent magnitude and color bounds. Image courtesy of M. Cooper of the Deep2 team.

3.12 Measures of Large scale Structure

This section arises because galaxies, especially massive ones, have a substantial affinity for each other; first they weakly cluster in a way that can be modeled linearly, but then clustering becomes non-linear, with significant quantities of kinetic energy deposited in coalescing systems. In this connection, two sub-topics are prominent – the correlation functions of galaxies, and the "bias" in galaxy formation relative to the ambient dark matter density. The signal of clusters is most trivially seen in histograms of galaxy redshift surveys with modest sky coverage. Figure 5 shows the redshift distribution of 8,412 galaxies in 1 square degree in one of the DEEP2 Survey fields. The histogram was smoothed. The error bars are 1-σ Poisson errors ($\pm\sqrt{N_{\rm bin}}$) were calculated according to the number of galaxies per bin for the smoothed distribution of galaxies. Thus, the peaks above the error bars show highly significant over-densities, and the valleys would probably represent voids. Voids of size $\delta z \simeq 0.8$ (four bins wide here) at $z \simeq 1$ would be ~ 90 Mpc in diameter (not comoving), for our standard parameters. Over-dense structures seem to be generally smaller. Let us look more closely at the more quantitative measures of galaxy clustering mentioned above.

3.12.1 The Correlation Functions

Astronomers often describe the quantitative tendency of galaxies to cluster by a two-point spatial correlation function $\xi(r)$, where r is the radial separation. Assume the galaxies have an average density n. With the distribution of galaxies tending toward clumpiness, the joint probability of locating a galaxy in one volume element, V_1, near another volume element, V_2, is described by a joint probability function of finding a galaxy in each volume simultaneously is:

$$\Delta P = n^2 \left[1 + \xi(r_{1,2})\right] V_1 V_2. \tag{5}$$

The function $\xi(r)$ represents the clustering of galaxies over and above the values found in a truly random (by random-number generator) catalog. Thus, the estimator for ξ was originally defined (Bahcall, 1986),

$$\xi(r) \equiv \frac{\langle DD(r) \rangle}{\langle RR(r) \rangle} - 1, \tag{6}$$

where r is the radial separation of the galaxies, $\langle DD \rangle$ is the observed mean frequency of galaxy pairs at distance r, and $\langle RR \rangle)$ is the mean frequency in random catalogs. Certain modifications to this estimator have been proposed with a minimal effect (Zehavi et al., 2002).

Empirically, the spatial correlation function has been found to be well-modeled by a power-law,

$$\xi(r) = \left(\frac{r}{r_0}\right)^{-\gamma}, \tag{7}$$

where r_0 is the correlation length.

The direct determination of ξ is problematic since radial distances must be inferred from the angular separations of galaxy pairs and their relative redshifts. However, at low redshifts the peculiar velocities of galaxies often overwhelm the Hubble flow, leading to the "finger of God" artifact*, which results when a cluster is sampled in a galaxy redshift survey.

Thus, it is the *angular* correlation function $w(\theta)$ which is more fundamental from an observational perspective. According to Landy & Szalay (1993), the optimal estimator of $w(\theta)$ is

$$w(\theta) = \frac{\langle DD\rangle - 2\langle DR\rangle + \langle RR\rangle}{\langle RR\rangle} - 1, \tag{8}$$

where $\langle DR\rangle$ is the pair mean frequency between one element of the galaxy catalog and one of the simulated catalog at angle $\theta \pm \mathrm{d}\theta/2$, and the angular dependency of the pair frequencies are suppressed for simplicity. If random catalogs have different numbers of simulated galaxies than the real catalogs, normalizing factors are added (Landy & Szalay, 1993). This analysis produces an empirical angular correlation of the form,

$$w(\theta) = A_w \left(\frac{\theta}{\theta_0}\right)^{-\delta}, \tag{9}$$

where A_w is the value of the correlation function at θ_0, and the exponent δ is found to be in the range of 0.7 or 0.8.

Because $w(\theta)$ is a projection of the $\xi(r)$, the slope of the angular correlation function may be calculated. Limber's equation (e.g., Limber 1954; Fall & Tremaine 1977) is the result of the statistical analysis that is used to express the projection of the spatial correlation function $\xi(r)$ into the angular correlation function $w(\theta)$. The inversion of the angular correlation function is a specialized technique that is beyond the scope of the present work, but with the knowledge that the spatial correlation function has a slope equal to one plus the angular correlation function δ, the procedure allows one to solve for the correlation length r_0.

The results of recent studies using SDSS data (e.g., Budavári et al., 2003; Zehavi et al., 2002) show that the correlation length is a function of galaxy luminosity, and therefore of mass. Local galaxies are found to have a mean correlation length $r_0 \simeq 5.8\ h^{-1}$ Mpc and a spatial correlation function slope $\gamma \simeq 1.8$. In addition, both the correlation length and the slope of the spatial correlation function are found to be a function of galaxy type. In this sense, red galaxies are found to have larger correlation lengths ($\sim 6.6\ h^{-1}$ Mpc) at a given luminosity, but a steeper correlation function than the mean ($\gamma \simeq 1.96$). Blue galaxies, however, have smaller correlation lengths ($\sim 4.5\ h^{-1}$ Mpc) and a flatter slope $\gamma = 1.68$.

The variation of correlation length with mass can be explained by a "universal" correlation function,

$$\xi_{\mathrm{i}} = 0.3(r/d_{\mathrm{i}})^{-\gamma}, \tag{10}$$

*In "pie" diagrams, clusters appear stretched in the radial direction, "pointing" toward the observer, as a result of the kinematic redshift distortions.

first proposed by Bahcall (1986),where $d_{\rm i}$ is the mean separation of $d_{\rm i} = n_{\rm i}^{-1/3}$ of objects of class mass-class i. Thus, clusters, which have a much smaller number density than galaxies, have a correspondingly larger correlation length. This relation is shown in Fig. 6, where the correlation length is shown as a function of SDSS absolute magnitudes.

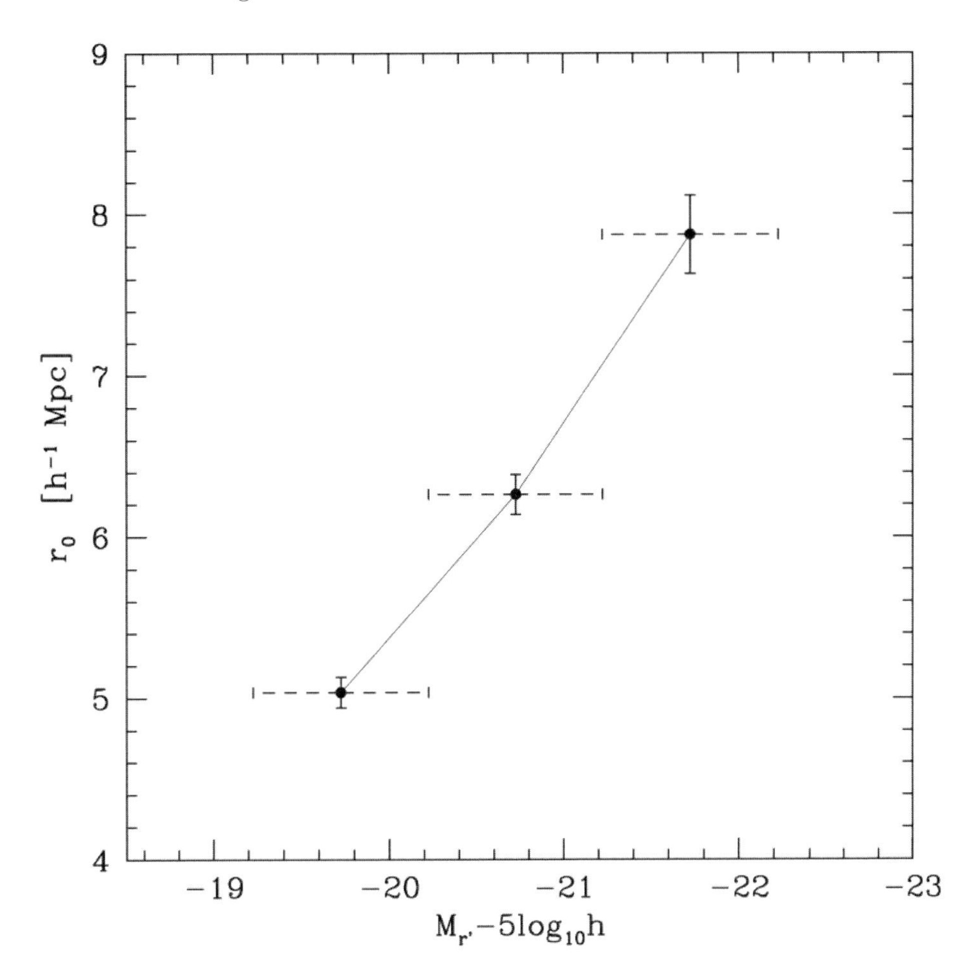

FIGURE 6: The correlation length of Sloan (SDSS) galaxies binned according to their absolute magnitude, showing that it scales with luminosity, and therefore, plausibly with mass. This form of relation is anticipated by the universal correlation function (see text). Plot is from Budavári et al. (2003).

Because the mean comoving density of galaxies or clusters is approximately constant with redshift, it is reasonable to suppose that high-redshift correlations may have comparable correlation lengths, when measured in comoving terms. However, the results of recent studies are quite variable. Both Coil et al. (2004) and Phleps & Meisenheimer (2003) find that clustering strength (the comoving correlation length) increases with time from values on order 1.5 h^{-1} Mpc at $z \simeq 0.85$ to the $\sim 5.5\, h^{-1}$ Mpc locally, though with a large uncertainty. Incidentally, Coil et al. (2004) found the same tendency, noted by Budavári et al., for red/blue objects to have larger/smaller correlation lengths and steeper/flatter slopes, respectively, at high redshift than the mean. The source of this bimodality, as we note below, may be the result of the variation of the mass-luminosity ratio with galaxy color.

Other observational papers suggest that comoving correlation lengths are relatively unaffected by increasing redshift. For instance, (Ouchi et al., 2004), who studied LBGs ($L \gtrsim L^*$) at $z \simeq 4$ and $z \simeq 5$, find large comoving correlation lengths of about 4.5 and 6 Mpc, respectively, suggesting a counter-trend to that noted by Coil et al. (2004) and Phleps & Meisenheimer (2003).

Correlation lengths can also be modeled from smoothed particle hydrodynamic (SPH) numerical simulations (Weinberg et al., 1999). In thi work, the comoving galaxy spatial correlation function was computed at redshifts ranging from $z = 0.5$ to $z = 4$. Results showed that the correlation length ($r_0 \simeq 5.2\ h^{-1}$ Mpc), and the slope ($\gamma \simeq 1.8$), should be essentially invariant with redshift (shown in the upper and lower panels, respectively, of Fig. 7). In this work, the mass of the galaxy is the relevant variable, while for the observational work mentioned above, it is the luminosity which is used. The discordant results of Coil et al. (2004) and Phleps & Meisenheimer (2003), relative to the modeling of Weinberg et al., may be explainable in terms of the lowering mass-luminosity ratios of galaxies with increased redshift (Conselice et al., 2005). We take up this issue again at the end of the next section.

3.12.2 The Question of Galaxy Biasing

What is *bias*? Bias is one of those theoretical quantities that exists because it can. It is a parameter that relates the distribution of baryons relative to dark matter. In practice, it is "measured" in a variety of ways, and so it has a variety of operational definitions, depending on the perspective of the researcher. At base, the bias conveys the ratio of the strength of matter fluctuations of baryons relative to those of the dark matter.

$$b \equiv \frac{\delta_{\mathrm{B}}}{\delta_{\mathrm{D}}}, \tag{11}$$

where δ is the over-density – the excess density relative to the mean. "B" refers to the distribution of baryons and "D" refers to dark matter. While the numerator is fairly straightforwardly determined from observations, it is much more difficult to extract dark matter (DM) fluctuations. In practice, the DM fluctuations are determined by the output of cosmological simulations; observationally, they are not well constrained.

The bias, as a function of time or redshift, may sensitively reflect the coupled evolution of baryons, dissipational particles, and dark matter, which is dissipationless. However, to calculate the bias, we need to know the nature of the DM fluctuations (see Eq. (11)). To do this, there must be a marriage of observational data and cosmological simulations. Put simply, a correspondence must be established between simulated baryonic galaxies and observed galaxies. Then the simulated dark halos can be assumed to represent the corresponding real dark halos, so that the simulated DM fluctuations can be taken to correspond to reality. Because the DM may be detected by its effect on galaxy rotation curves, it is possible to check on the quality of the simulations by extracting the details of simulated galaxies such as the variation of bias with radius, mass, and age of galaxies.

We think the early cosmic history of bias must be somewhat interesting. Before recombination, DM was able to cluster, but baryon clustering was suppressed by their being coupled to the radiation field. Following decoupling, baryons could effectively cool, and fall into the potential wells of waiting DM halos. The effect of this must have been to enhance the central condensation of DM halos somewhat. But as baryons continued to cool, galaxies were born, and the feedback of heightened luminosity and kinematic outflows served to restrain the pace of consolidation of luminous systems. More recently, other complex forces may be at work in the more massive agglomerations of matter. For instance, Kravtsov & Klypin (1999) suggest that dynamical friction and tidal stripping in cluster environments may be responsible for reducing the number density of baryon "halos" in clusters and groups, resulting in a reduced bias, even "anti-bias" (i.e., $b < 1$) at $z \lesssim 0.5$ for larger over-densities.

Bias is calculated in a couple of ways other than that of Eq. (11). One method is by the ratio of the correlation length of baryonic matter to that of dark matter:

$$b_{\xi}(r) = \sqrt{\frac{\xi_{\mathrm{B}}(r)}{\xi_{\mathrm{D}}(r)}}. \tag{12}$$

As measured in simulations, the bias is generally a mild function of radius. Beyond distances of $\sim$300 kpc (close to the virial radius), the bias is seen to increase gradually with distance, and within that distance, it also increases, in response to dissipational clustering of baryons (Weinberg et al., 2004).

A third method of bias estimation exists, which is based on the power spectra $P(k)$ of the baryons and the dark matter,

$$b_{\mathrm{P}} = \sqrt{\frac{P_{\mathrm{B}}(k)}{P_{\mathrm{D}}(k)}}, \tag{13}$$

where P is the power spectrum, and k is the wave number. Each of these three definitions are in some sense different, for since correlations, power, and mass fluctuations sample the same variable field, but in different ways.

We might note, at this juncture, that Eq. (11) is obviously sensitive to the mass of the system. Equation (12), if it is to be consistent with Eq. (11), should use a correlation function of objects that is based on their masses rather than their luminosities. Of course this is observationally more difficult!

In general terms, these measures of bias all probe the same trends in the evolution of structure. Figure 7 shows the evolution of the comoving correlation lengths and correlation function slope for the baryonic part of simulated galaxies, and for

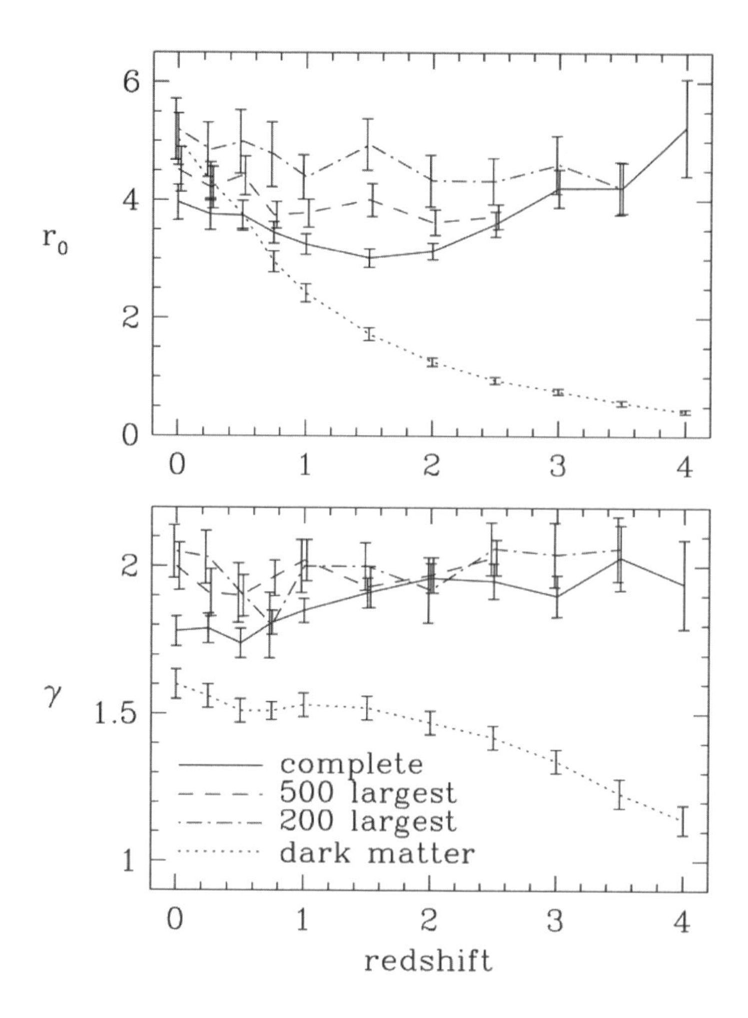

FIGURE 7: Upper panel, the comoving correlation lengths for simulated galaxies (the complete catalog, 500, and 200 largest; solid, dashed, dot-dashed, respectively), and for DM halos (dotted). As discussed in the text, DM experiences no moderation in the growth of the correlation scale with time. Lower panel, the evolution of correlation function slope γ for simulated galaxies (as in upper panel) and dark matter. Figure from Weinberg et al. (2004).

the DM halos (Weinberg et al., 2004). Note that while the model baryonic systems display a relatively constant trend in the correlation length, the modeled dark halos display a growing correlation length as a result of progressive hierarchical clustering. The coalescence of DM is also shown in the temporally increasing slope of the correlation function, γ. Taking the ratio of the correlations according to Eq. (12), we see that bias must evolve rapidly with redshift, declining from a value near 5 or so at high redshift, to a value near unity at present. However, these simulation results are at variance with Coil et al. (2004) and Phleps & Meisenheimer (2003), who observe a decreasing correlation length as one looks to higher redshift, and therefore expect a bias that is much-less variable with redshift. Who should we believe?

A possible problem in attempting to observe the evolution of the bias with redshift or look-back time can be seen by inspection of Eq. (11). It is the galaxy mass which is relevant to its calculation, not the luminosity, as noted at the end of §3.12.1. Observational studies tend to use magnitude-limited samples, which are not sensitive to changes in the specific SFR and mass-luminosity ratio. Kauffmann et al. (2003) and Conselice et al. (2004) study morphology and mass-luminosity ratios of galaxies in the HDFN & S, by finding a best-fit history to the SED, thereby enabling them to determine the stellar M/L. Their results show a broad range of M/L values at low redshift, but almost uniformly low values ($\sim 0.2^{+0.2}_{-0.1}$) for $z \gtrsim 1.8$. Such a trend with redshift is to be expected since the star formation rate density is known to have been significantly greater at $z \gtrsim 0.5$ (Lilly et al., 1995; Hippelein et al., 2003). Thus, an absolute magnitude limited survey will pick lower-mass galaxies at high redshift than those considered at low-redshift. Low-mass halos at high redshift will have smaller correlation lengths. This trend may substantially explain the discrepancy between observed correlation lengths and modeled correlation lengths.

3.13 Summary

We have reviewed some of the main aspects of the study of the evolution of the Universe. We have found that galaxy counts are not well-explained by a non-evolving population, as a type-dependent excess over the predictions of non-evolving models manifests itself at faint magnitudes; blue galaxies have a large excess, while red galaxies have a moderate excess. The galaxies which contribute to this excess appear to be at redshifts on order $z = 1$, and have been a major subject of study for the last 25 years.

We found that as we looked back in time, the morphology of galaxies changed. We found that the Hubble sequence progressively ceases to be relevant as one probes higher redshifts. Two new kinds of objects appear – the LAOs (high asymmetry), and the LDOs (low concentration) – which combine to host about one-third of the total SF in galaxies at $1 \leq z \leq 2$. We found that the indices of asymmetry and clumpiness were presently not very useful for the analysis of morphology at $z \gtrsim 1$ (they are surface-brightness dependent), but the concentration index remains quite useful to high redshift, where the data suggests generally lower concentration indices.

In the currently standard hierarchical model of structure formation, the most massive objects collapse last. These would be the galaxy clusters. Therefore, we expected to be able to detect the evolution of clustering over the last several Gyr. Rich clusters are easily detected at moderate redshifts with X-ray space telescopes, as their surface brightness is a function of their virial temperature. However, at high redshifts, a surface brightness problem sets in; a given bolometric surface brightness declines as $(1+z)^{-4}$. But even so, there is little evidence of evolution of the cluster LF out to $z \sim 0.8$, except perhaps a shortage of the most luminous ($L \gtrsim 10^{44}$ erg s^{-1}). When the new programs to detect the SZE-effect in clusters are in place, we expect to be able to robustly sample the cluster mass function out to high redshift as a result of the negative K-corrections in the CMB microwave region.

The global SFR as a function of redshift shows a dramatic (~10-fold) increase between the present and $z = 1$. Emission lines such as [O II] or [O III] may be used to sample the SFR out to $z \sim 1$. The evolution of the SFR density at higher redshifts remains somewhat murky because of the problems in correcting for extinction and accounting for the undetected galaxies too faint to observe. Recent results suggest a gradually diminishing comoving SFR density for $z \gtrsim 1.5$.

The expected increase in metallicity with time, while seemingly incontrovertible, has few clear and unambiguous signals. A variety of targets, from clouds to DLAs, to galaxies and AGN, offer themselves, each presenting unique problems of interpretation. The R_{23} estimator is useful, but its application is burdened with an ambiguity that requires an additional assumption, or information, to be useful for metallicities slightly sub-solar. The conclusion which best fits the data is that the metallicity, and the rate of increase of metallicity, is strongly correlated to the density of the environment. The bottom line on this subject is that the metallicities of "ancient" Galactic globular clusters are far below the metallicities of stars forming today in the Galaxy. However, some few number of dwarf galaxies are forming stars of lower metallicity than many Galactic globular clusters. Dwarf star-formers, like II Zw 40 and I Zw 18, are good metal-poor examples.

The evolution of AGN is first sampled by the traditional $\langle V/V_{\rm max}\rangle$ test, which shows higher than expected numbers for surveys at $z \lesssim 2$, but a decline for $z \gtrsim 2.5$. The luminosity function of AGN shows clear luminosity evolution, but there may also be some number evolution. The best models of the X-ray LF are double power-laws, and redshift-binned LFs show that the characteristic luminosity (at the "knee" of the LF) increases strongly with increasing redshift. Work with Chandra has enabled us to find the obscured AGN, which suggests a gradually increasing population of Seyfert-type AGN, dominating the XLF at low redshift. This anti-hierarchical AGN formation may be governed by a more effective feedback when the potential well is shallow.

Correlation functions embody a more statistical approach to large-scale structure. Together with simulation-aided calculation of the bias, these relations may be used to trouble-shoot modern galaxy and large-scale structure simulations.

Detecting the enrichment of the cosmos by following the evolving metallicities of various classes of astrophysical objects (see Table 2) show little definite evolution. What needs to happen is to be able to follow a given astrophysical object in time, and watch its evolution from cloud to pre-galactic to galactic or QSO. This fine ability, however, is not open to us. We must either content ourselves by models of the production of metals, or simply to measure the mean metallicity of the Milky Way's stars as a function of their age. Neither are the direct path we'd like to follow to understand metallicity as a function of cosmic epoch.

REFERENCES

Abraham, R. G., Merrifield, M. R., Ellis, R. S., Tanvir, N. R., & Brinchmann, J., et al., 1999, MNRAS, 308, 569

Abraham, R. G., et al., 1996, ApJS, 107, 1

Allende Prieto, C., et al., 2001, ApJL, 556, L63

Babul, A. & Rees, M. J. 1992, MNRAS, 255, 346

Babul, A, Ferguson, H. C. 1996, ApJ, 458, 100

Bahcall, N. A. 1977, ARAA, 15, 505

Bahcall, N. A. 1986, ApJL, 302, L41

Baldwin, J. A., et al., 2003, ApJ, 583, 649

Barger, A. J., et al., 2003, ApJL, 584, L61

Bell, E. F., et al., 2004, ApJ, 608, 752

Bentz, M. C. & Osmer, P. S. 2004, AJ, 600, 000

Bershady, M. A., et al., 2000, AJ, 119, 2645

Bouwens, R. J., et al., 2004, ApJL, 611, L1

Bouwens, R. J., et al., 1997, ApJL, 489, L21

Boyle, B. J., et al., 2000, MNRAS, 317, 1014

Böhringer, H., et al., 2002, ApJ, 566, 93

Brinchmann, J., et al., 1998, ApJ, 499, 112

Bruzual A., G. 1983, ApJs, 53, 497

Bruzual A., G. 2001, Ap & SS Supp, 277, 221

Bruzual A., G. & Charlot, S. 1993, ApJ, 405, 538

Budavári, T., et al., 2003, ApJ, 595, 59

Bunker, A. J., Spinrad, H., Stern, D., Thompson, R., Moustakas, L., Davis, M., & Dey, A. 2000, in *Galaxies in the Young Universe II*, eds. H. Hippelein & K. Meisenheimer, 2000; astro-ph/0004348

Carlstrom, J. E., et al., 2002, ARAA, 40, 643

Cattaneo, A. & Bernardi, M. 2003, MNRAS, 344, 45

Cayón, L., et al., 1996, ApJL, 467, L53

Cen, R. & Ostriker, J. P. 1999, ApJL, 519, L109

Cimatti, A., et al., 2004, Nature, 430, 184

Cohen, S. H., et al., 2003, AJ, 125, 1762

Coil, A. L., et al., 2004, ApJ, 609, 525

Conselice, C. J., et al., 2000a, AAp, 354, L21

Conselice, C. J., et al., 2000b, ApJ, 529, 886

Conselice, C. J., et al., 2000b, AJ, 119, 79

Conselice, C. J., et al., 2003, ApJL, 596, L5

Conselice, C. J. 2003, ApJS, 147, 1

Conselice, C. J., et al., 2004, ApJL, 600, L139

Conselice, C. J., et al., 2005, ApJ, 620, 564

Cowie, L. L., et al., 2003, ApJL, L584, L57

Cowie, L. L., et al., 1995a, AJ, 110, 1576

Cowie, L. L., Songaila, A., Kim, T., & Hu, E. M. 1995, AJ, 109, 1522

Croom, S. M., et al., 2004, MNRAS, 349, 1397

Daddi, E., et al., 2004, ApJ, 617, 746

Dietrich, M., et al., 2002, ApJ, 564, 581

Dietrich, M., et al., 2003, ApJ, 596, 817

Drory, N., et al., 2004, ApJ, 608, 742

Dunlop, J. S. 1999, in ASP Conf. Ser. 193: *The Hy-Redshift Universe: Galaxy Formation and Evolution at High Redshift*, p. 133

Dunlop, J. S., et al., 1996, Nature, 381, 581

Ebeling, H., et al., 2000, ApJ, 534, 133

Edmunds, M. G. 1999, in ASP Conf. Ser. 187: *The Evolution of Galaxies on Cosmological Timescales*, 203–213

Ellis, R. S. 1997, ARAA, 35, 389

Fall, S. M. & Tremaine, S. 1977, ApJ, 216, 682

Ferguson, H. C. & Babul, A. 1998, MNRAS, 296, 585

Ferguson, H. C., et al., 2004, ApJL, 600, L107

Flores, H., et al., 1999, ApJ, 517, 148

Folkes, S., et al., 1999, MNRAS, 308, 459

Fried, J. W., et al., 2001, AAp, 367, 788

Giavalisco, M., and 56 co-authors, 2004, ApJL, 600, L93

Glazebrook, K., et al., 1999, MNRAS, 306, 843

Granato, G. L., et al., 2001, MNRAS, 324, 757

Guzman, R., et al., 1996, ApJL, 460, L5

Haehnelt, M. G., et al., 1998, MNRAS, 300, 817

Haiman, Z. & Hui, L. 2001, ApJ, 547, 27

Hamann, F. & Ferland, G. 1999, ARAA, 37, 487

Heckman, T. M., et al., 1998, ApJ, 503, 646

Hippelein, H., et al., 2003, AAp, 402, 65

Hogg, D. W. et al. 1998, ApJ, 504, 622

Joy, M., et al., 2001, ApJL, 551, L1

Kauffmann, G., et al., 2003, MNRAS, 341, 33

Kaviani, A., et al., 2003, MNRAS, 340, 739

Kennicutt, R. C. 1992, ApJL, 388, 310

Kennicutt, R. C., et al., 1994, ApJ, 435, 22

Kolatt, T. S., et al., 1999, ApJL, 523, L109

Koo, D. C., et al., 1994, ApJL, 427, L9

Koo, D. C. & Kron, R. G. 1992, ARAA, 30, 613

Kravtsov, A. V. & Klypin, A. A. 1999, ApJ, 520, 437

Kuraszkiewicz, J. K. & Green, P. J. 2002, ApJL, 581, L77

Landy, S. D. & Szalay, A. S. 1993, ApJ, 412, 64

Laor, A., et al., 1994, ApJ, 420, 110

Lilly, S. J., et al., 2003, ApJ, 597, 730

Lilly, S. J., et al., 1995, ApJ, 455, 108

Limber, D. N. 1954, ApJ, 119, 655

Lin, D. N. C. & Murray, S. D. 1992, ApJ, 394, 523

Longair, M. S. 1998, *Galaxy Formation* (Berlin: Springer)

Madau, P., et al., 1996, MNRAS, 283, 1388

Madau, P., et al., 1998, ApJ, 498, 106

Magorrian, J., et al., 1998, AJ, 115, 2285

Manning, C., et al., 2000, ApJ, 537, 65

Martin, C. L. & Sawicki, M. 2004, ApJ, 603, 414

Marzke, R. O., et al., 1998, ApJ, 503, 617

Marzke, R. O., et al., 1994, AJ, 108, 437

McCarthy, P. J. 2004, ARAA, 42, 477

McGaugh, S. S. 1991, ApJ, 380, 140

Möller, P., et al., 2004, AAp, 422, L33

Murray, S. D. & Lin, D. N. C. 1992, ApJ, 400, 265

Nolan, L. A., et al., 2001, MNRAS, 323, 385

Oke, J. B. & Korycansky, D. G. 1982, ApJ, 255, 11

Ouchi, M., et al., 2004, apJ, 611, 685

Pagel, B. E. J., et al., 1979, MNRAS, 189, 95

Patton, D. R., et al., 2002, ApJ, 565, 208

Peterson, B. M. 1997, *An Introduction to Active Galactic Nuclei*, p. 128 (Cambridge: Cambridge University Press)

Petrosian, V. 1976, ApJL, 209, L1

Pettini, M., et al., 2002, AAp, 391, 21

Phleps, S. & Meisenheimer, K. 2003, AAp, 407, 855

Press, W. H. & Schechter, P. 1974, ApJ, 187, 425

Prochaska, J. X., et al., 2003, ApJL, 595, L9

Rix, H. & Zaritsky, D. 1995, ApJ, 447, 82

Rosa-González, D., et al., 2002, MNRAS, 332, 283

Rosati, P., et al., 1999, AJ, 118, 76

Rosati, P., et al., 2002a, ARAA, 40, 539

Rosati, P., et al., 2002b, ApJ, 566, 667

Salucci, P., et al., 2000, MNRAS, 317, 488

Schmidt, M. 1968, ApJ 151, 393

Schmidt, M., et al., 1988, in ASP Conf. Ser. 2: *Optical Surveys for Quasars*, 87

Schmidt, M., et al., 1991, in ASP Conf. Ser. 21: *The Space Distribution of Quasars*, 109–114

Schneider, D. P., et al., 1989, AJ, 98, 1507

Schreiber, N. M. F., et al., 2004, ApJ, 616, 40

Sheth, K., et al., 2003, ApJL, 592, L13

Shields, G. A., et al., 2003, ApJ, 583, 124

Shu, F. H. 1977, ApJ, 214, 488

Spinrad, H., et al., 1997, ApJ, 484, 581

Stanford, S. A., et al., 1997, AJ, 114, 2232

Stanford, S. A., et al., 2001, ApJ, 552, 504

Stanford, S. A., et al., 2002, ApJS, 142, 153

Steffen, A. T., et al., 2003, ApJL, 596, L23

Steidel, C. C., et al., 1999, ApJ, 519, 1

Stern, D., et al., 2003, AJ, 125, 2759

Stockton, A. & Ridgway, S. E. 1998, AJ, 115, 1340

Sunyaev, R. A. & Zeldovich, Y. B. 1969, Nature, 223, 721

van den Bergh, S., et al., 2000, AJ, 120, 2190

van den Bergh, S., et al., 2002, AJ, 123, 2913

Vladilo, G. 1999, in ASP Conf. Ser. 187: *The Evolution of Galaxies on Cosmological Timescales*, 323–337

Vladilo, G., et al., 2000, ApJ, 543, 24

Warren, S. J., et al., 1988, in ASP Conf. Ser. 2: *Optical Surveys for Quasars*, 96

Wechsler, R. H., et al., 2002, ApJ, 568, 52

Weedman, D. W. 1986, *Quasar astronomy* (Cambridge and New York: Cambridge University Press, p. 226)

Weinberg, D. H., et al., 2004, ApJ, 601, 1

Weinberg, D. H., et al., 1999, in ASP Conf. Ser. 191: *Photometric Redshifts and High Redshift Galaxies*, 341

Weinberg, S. 1972 *Gravitation and Cosmology*, John Wiley & Sons

Williams, R. E., et al., 1996, AJ, 112, 1335

Wills, D. & Lynds, R. 1978, ApJS, 36, 317

Zaritsky, D. 1995, ApJL, 448, L17

Zaritsky, D. & Rix, H. 1997, ApJ, 477, 118

Zehavi, I., & 66 co-authors 2002, ApJ, 571, 172

Chapter 4

Galaxies at the Contemporary Limits

Limited by both our knowledge and the availability of modern telescopic and detector technology we'll probe frontier research on the most-distant (and youngest) galaxies that can be observed with contemporary techniques. We can, with difficulty, obtain limited observational data on systems up to look-back times $\sim 90\%$ the expansion age of the Universe.

4.1 Non-traditional Searches for Great Distance

The surface density of faint galaxies on the sky plane is imposing. To an I (814 nm) magnitude of 26.5 there are $\sim$75 galaxies (cumulative count) per square arcmin ($\square'$) detected at high galactic latitudes. For more on the faint "raw galaxy counts" in different wave-bands, please refer to the HST data of Williams et al. (1996).

Clearly, to study the few faint galaxies which are really at a large redshift (roughly 3% with $I_{\rm AB} \leq 26.5$ at $z > 5$) we need to discriminate them robustly; this can be accomplished in several ways – with varying efficiencies and, not surprisingly, with the introduction of observational biases. For two previous discussions along these lines, see Stern et al. (1999) and Spinrad (2004).

We divide the available selection techniques into "traditional" approaches and "non-traditional" approaches. We treat the latter first to compensate for their less-prominent status. For, as we shall see, the following non-traditional sources may provide unique perspectives on high-redshift objects.

4.1.1 Radio Surveys

Non-traditional wave-bands (i.e., non-optical) can yield significant "extra" information to aid in our identification of a distant galaxy. For example, distant radio sources (mainly E systems with fairly steep radio synchrotron spectral indices) can be signposts for the selection of distant large galaxies and their clustered neighbors. Radio-loud galaxies at $z \geq 2$ often emit radio spectra steeper than the "normal" $\alpha \simeq 0.7$, where the exponent α in

$$f_\nu = K\nu^{-\alpha} \tag{1}$$

represents the "steepness" of the radio spectrum. Nominal steep spectra have $\alpha > 1.0$ (van Breugel et al., 1999; Chambers et al., 1990). With some experience astronomers have also learned about which range of radio flux-densities optimizes the search for distant sources. We often wish to discriminate between faint but intrinsically powerful QSRs and nearby, intrinsically lower-luminosity radio galaxies. The steepness of the radio spectrum allows us to do this, although not perfectly.

In the bright Revised Third-Cambridge Radio Source catalogue (hereafter, the "3CR" catalogue), the maximum radio luminosity is near $\log P_{408} = 29.6$ (W Hz^{-1}) at 408 MHz (McCarthy, 1999) for radio galaxies at $z \geq 1.8$. The highest redshift in the bright-source 3CR is that of the galaxy 3C 257 (Stern & Spinrad, 1999) at $z = 2.474$. To penetrate the foreground galaxy "clutter" needed for the location of really distant radio galaxies (say, at $z > 4$), we would imagine searching for sources apparently fainter than the 3CR limit by perhaps a factor of 5 to 20. Such steep sources (in the mid-radio region of $\sim$ 1.4 GHz), with flux densities near 100 mJy and spectral indices $\alpha \geq 1.3$ are now recognizable in modest numbers. And a few faint radio galaxies have turned out to be at very high redshift (see van Breugel et al. 1999 and De Breuck et al. 2000). For example, following van Breugel et al. (1999), we list in Table 1 some steep spectrum radio sources associated with galaxies at $z > 4$. We note parenthetically that very few radio-loud QSRs (quasars) have been discovered at $z > 4$ and perhaps there are only one or two known at $z > 5$.

TABLE 1: Distant Radio Galaxies

Name	z	α_{365}^{1400}	Reference
TNJ0924-2201	5.19	1.63	van Breugel et al. (1999)
VLAJ1236+6213	4.42	0.96	Waddington et al. (1999)
6C 0140+326	4.41	1.15	De Breuck et al. (2000)
8C 1435+63	4.25	1.31	Spinrad et al. (1995)
TNJ 1338-1942	4.11	1.31	De Breuck et al. (1999) and Venemans et al. (2002)

Note: This α is the power-law index for the radio spectrum in the range 365 to 1400 MHz.

The radio galaxies in Table 1 are intrinsically powerful; some or most are associated with luminous galaxies – and are likely already (as seen by us) fairly massive systems that must have started their accumulation of dark matter, baryons, and probably an MBH very early in their evolutionary path.

Another valuable property of the radio galaxies, as may also be the trend with quite large E systems, is their tendency to be located in relative dense environments. Their spatial correlation function has a large amplitude. So groups and clusters can be located near radio galaxies, even at redshifts $z \geq 4.0$. Thus the result by Venemans et al. (2002) for the galaxy TN1338-1942 (see Table 1); imaging and spectroscopy with the VLT (i.e., Very Large Telescope) revealed ~20 Lyα emitters with the same ($z \simeq 4.1$) redshift as the radio galaxy – in this over-dense (by a factor of about 15) region. The projected size scale of the dense region is at least 2.7×1.8 Mpc (comoving), with a velocity dispersion of 325 km s^{-1}.

4.1.2 Surveys at Energetic Wave-bands

Other non-optical means of selecting distant galaxies use the energetic side of the electromagnetic spectrum. We are just starting to recognize that gamma-ray bursts (GRB) are energetic phenomena related to SNe, and luminous enough to be briefly detectable (by their "afterglows", really) at cosmological distances. Some of the GRBs that are satisfactorily located and with robust spectroscopy are quite distant, $z \geq 2$. To be more quantitative, five out of 28 GRBs with redshifts are at $z > 2$ (Bloom, 2003)!

The GRBs host galaxies, when successfully identified, are usually (always?) star-formers. Djorgovski et al. (2003) point out a GRB host with the characteristics of a starburst at $z = 1.10$. The link between GRB/SN explosions and a host galaxy forming massive stars at a high rate is quite provocative, and indirectly pushes our assessment of the event toward massive stellar evolution.

Can GRBs take us to the very distant Universe, $z \geq 5$? Andersen et al. (2000) have reported a GRB-optical afterglow spectrum with a sharp continuum break near λ6710 Å; if interpreted as the onset of the Lyα forest absorption we would then suggest a redshift of $z \simeq 4.50$. Still at magnitude $\gtrsim 26$ and with Lyα in the near IR, such briefly available targets at $z > 5$ will be a real observational challenge for 8–10 m class telescopes. On the other hand, the resulting spectroscopic analysis could be straightforward. We leave this futuristic method of locating distant GRBs and their host galaxies in this promising, but unproven situation.

Another high-energy selection for possibly locating very distant galaxies with energetic photons uses the capability of the Chandra satellite for X-ray observations at a range of energies. The nuclei of mild AGN and stronger ones (QSOs in the extreme) emit X-rays at luminosities that allow their detection at cosmological distances by virtue of long Chandra integration times. These emissions are probably related to the emissions from the environs of MBH. Several X-ray and optical surveys have now matured to the point of becoming a realistic method to locate distant X-ray galaxies, if they exist in substantial numbers. The new Chandra surveys disclose a population of apparently faint but X-ray-luminous galaxies at $z > 4$ (Barger et al., 2002). Applying the Lyman break technique to the optical identifications with the Subaru prime-focus camera reveals a population of optically faint sources that could potentially lie at $z \gtrsim 5$. Relative to projections of the QSO luminosity function out to $z \simeq 4.5$ (Haiman & Loeb, 1998), this population is a factor of 50 smaller than expectations, according to Barger et al. (2003), who show that $z > 5$ AGN cannot have been responsible for the reionization of the Universe. Contrary to our desire to search successfully for extremely distant and perhaps unusual galaxies, Barger et al. (2003b) find for their hard X-ray sources 54% originate in host systems at $z < 1$, and 68% of their global luminosity comes from $z < 2$. The highest individual redshifts actually measured are $z = 3.9$, $z = 4.14$, and $z = 5.19$ (out of ~270 with spectra). So it appears as if the median redshift is near 1 for faint X-ray selected galaxies, likely to represent the Seyfert-type (either types 1 or 2) mild AGN. Very recently, Bremer et al. (2004) used HST ACS images of the Chandra 2 megasecond image finding that for $z > 5$ candidates in the ACS, none corresponded to Chandra sources, and thus concluded that luminous AGN could make only a small contribution to the final stages of re-ionization of the Universe.

4.1.3 Millimeter and Sub-millimeter Galaxies at the Limit

The advances in the instrumentation dealing with millimeter and sub-millimeter sources in the 1990s and early 2000s are beginning to bear valuable "fruit". While these systems fail to produce high-resolution maps, they are being improved in sensitivity, so that detections of distant sub-millimeter sources (the so-called ultra-luminous infrared galaxies, or ULIRGs) are now fairly robust.

The sub-millimeter wave SCUBA (Submillimeter Common User Bolometer Array) on the JCMT (James Clerk Maxwell Telescope in Hawaii) operating at 850 μm and the MAMBO (Max-Planck Millimeter Bolometer), an array operating at 1.2 mm (Kreysa et al., 1998) on the IRAM 30-m telescope (Instituto Radioastronomia Millimetrica in Spain) have been the instruments of choice in locating sub-millimeter and millimeter IR-bright galaxies since ~ 1997. However the SCUBA point-spread-function (PSF) is coarse compared to extragalactic resolution "standards"; we have had to content ourselves with surveys conducted with $\geq 10''$ beams and long integration times.

Luckily there is a fairly strong correlation between high-redshift sub-millimeter sources (the rest-frame IR), and those whose radio emission near 20 cm ($\nu = 1.4$ GHz) is strong enough to often pose as an identification surrogate. Radio continuum identification is a useful empirical method to improve the source positional estimates so that either an optical redshift or a radio (millimeter) region CO-line detection may (with difficulty) be obtained.

Chapman et al. (2003) have reviewed and summarized some of the most-recent and most-reliable spectroscopic redshifts (in the optical region) for ~50 radio-identified sub-millimeter galaxies. Another three to five galaxies were identified and their relevant CO transitions observed with IRAM.

Chapman illustrated the redshift distribution histogram for radio-identified SCUBA galaxies; the median redshift lies near $z = 2.4$ and the largest redshift to date is at $z = 3.7$. What is our easiest interpretation of the sub-millimeter galaxies? Though they have a surface density ~ 10% of that of Lyman break galaxies (LBGs) at $z \sim 3$, their SFR density is equal to that of LBGs (Chapman et al., 2005), meaning that individually, their SFR is $\sim$ 10 times that typical of LBGs.

The spectroscopy of the SCUBA sub-millimeter sources just mentioned has proven difficult; the galaxies are optically faint and, as noted by Chapman et al. (2003), the necessary UV spectral features (rather important at $z \simeq 2$ to 3) were only well detected after the new LRIS-B ultraviolet spectrograph became available at the Keck I telescope (Steidel et al., 2003). At the highest redshifts, $z \simeq 3.7$, the typical inferred SFRs are at least 100 $\mathrm{M_\odot\,yr^{-1}}$, *if* the source of dust-heating is starlight from a starburst. It is also possible (and in a few cases, likely) that an AGN is the main heating source (as some spectroscopic evidence shows the N V $\lambda_0$1240 Å or a C IV $\lambda_0$1549 Å line which requires an ionization source more energetic than normal OB stars). Perhaps $\sim$20% of the SCUBA galaxies appear unusually bright at radio wavelengths, again suggesting an AGN is likely in the sub-millimeter sample.

The morphologies of some SCUBA galaxies have recently been explored by imaging with the HST. Chapman et al. (2003) and Conselice et al. (2003) note these IR-galaxies often appear as multi-component and/or disturbed mergers caught in "their act". If they are not predominantly excited by AGN, they are likely gas-rich young building blocks for eventual coalescence and perhaps spheroid formation (Smail et al., 1991).

Can we push these IR-galaxies to and beyond the largest redshifts available to other galaxies found at shorter wavelengths? Probably not with SCUBA. However, in Chapter 6 to follow we consider the application of powerful new tools, especially the Atacama Large Millimeter Array (ALMA), in our search for very distant, dusty galaxies. The present-day SCUBA limits are a few ($\sim$3) millijansky at $\lambda = 850\,\mu$m; ALMA with its planned 64 12-m antennae should be more sensitive and less-confused (because of the interferometric nature of the concluding signal). Thus, many authors (see Omont et al. 2003) are optimistic that the ALMA can locate (and evaluate SFRs of) dusty galaxies $\sim$50 times fainter (at 850 μm) than those observed with SCUBA. It may reach past $z = 15$, if the properties of these young galaxies can be scaled from the conclusions reached at $z \sim 3$ millimeter data.

4.2 Traditional Searches for Distant Objects

There are at least three rather general and traditional sampling techniques that could be selectively useful in choosing "unusual galaxies", perhaps (or hopefully) those seen at the largest redshifts. These are, first, continuum measures, such as the Lyman break method, second, narrow-band photometry, which can be used to find emission-line galaxies, and third, serendipitous discoveries made from other data collected in traditional observing programs.

4.2.1 Pre-selections on the Continuum Shape: Galaxies at $z = 3$ to $z = 6.5$

With a rather general method, we often use the photometric colors and/or the spectral energy distribution (SED) of distant galaxies to pre-select special redshift ranges. If broad-band spectra have high-amplitude features, like the redshifted Lyman discontinuity, or the Lyα "forest" adsorption below $\lambda_0 1216$ Å, their "peculiar" continua can be utilized as robust redshift indicators. At lower redshifts the 4000 Å break has served the same purpose, especially for E galaxies with $z \leq 1.0$.

The idea of using broad-band colors to estimate redshifts has a long history, dating back to Baum (1962), who used nine-band photoelectric data to estimate galaxy cluster redshifts. Koo (1985) showed that contours of constant redshift on color–color plots may provide reliable photometric redshift estimates. These techniques have enjoyed a recent revival, catalyzed by the deep photometry of the Hubble Deep Field (HDF) and the new generation of large-format CCD arrays.

The practical utilization of good digital photometry, including UV wavelengths on large numbers of faint galaxies led Steidel & Hamilton (1992) and then Steidel et al. (1996) to select a number of U-drop ($\sim$3650$\pm$700 Å) galaxies. As anticipated for galaxies at redshifts $z \sim 3$ and greater, the Lyman limit intrinsic to the galaxy and to the Lyα forest of the IGM combine to greatly reduce the photon transmission below $\lambda_0 1216$ Å. Both of these processes are ubiquitous in extragalactic continua; they make the selection of star-forming systems particularly straightforward in two-color diagrams, where the short-wave depression (the U-drop) is severe. In searches for high-redshift LBGs, color-selection criteria impose the first cut in assembling a catalog. Figure 1 shows the color-selection criteria used by Steidel et al. (2003) for $z \sim 3$ galaxies; the U-drops lie in the trapezoidal region bounded by the lower solid line.

The Lyman-break technique has been pushed to longer wavelengths and higher redshifts. For instance, Steidel et al. (1999) use G-drops (a blue-green filter) to find $z \simeq 4.0$ LBGs, and the Great Observatories Origins Deep Survey (GOODS) team use V- and I-drops to find galaxies up to $z \sim 5.8$ (Giavalisco et al., 2004).

Shown in Figs. 2 and 3 are two examples of HST broad-band imaging of $z \simeq 5.19$ galaxies. While the Lyα emission line and the discontinuity at 1216 Å fall at $\lambda \sim 7500$ Å in the observer's frame, the Lyman limit falls at $\sim$5600 Å. Colors can be derived from such images, and basic photometric redshifts can be extracted. Coincidentally, the galaxies featured in both figures are are members of the same apparent proto-cluster at $z \simeq 5.2$.

The process of color selection to achieve a catalog of galaxies in a particular redshift range can be generalized into routines which sample the SED of a galaxy to approximate its redshift. These are known as photometric redshifts. Table 2 shows the rough redshift range that is attainable by using the Lyman break technique in the various band passes utilized by the GOODS survey (Giavalisco et al., 2004). Among the other, more sophisticated techniques for photometric redshifts are Bayesian approaches, template-fitting methods (using synthetic or empirical

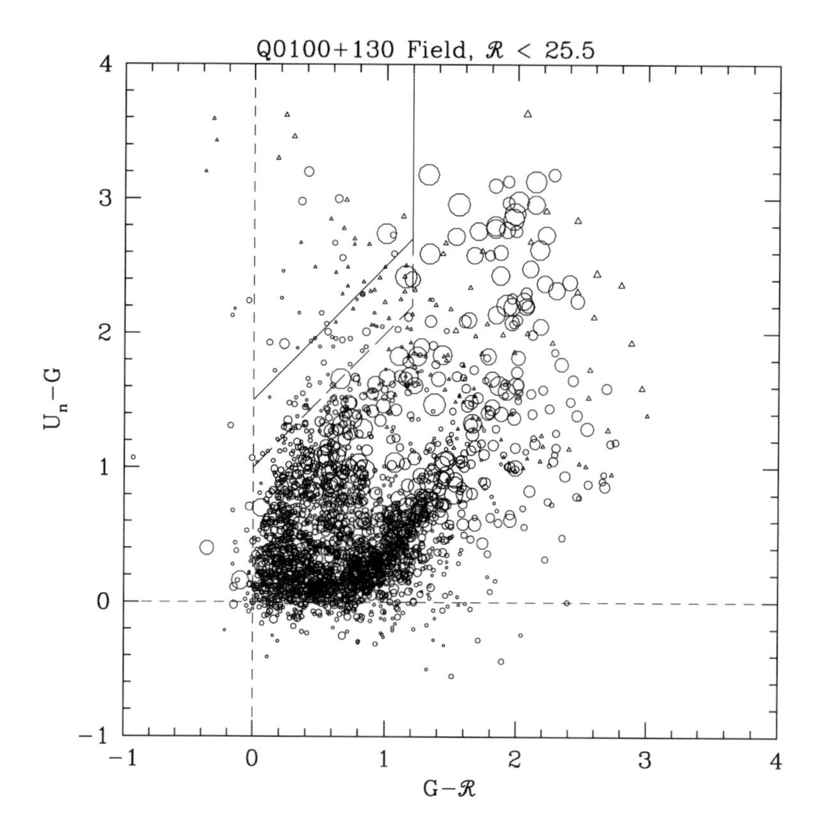

FIGURE 1: A color–color diagram in which data is overplotted on color space so that galaxies with colors consistent with $z \sim 3$ galaxies can be selected. The upper trapezoidal region safely avoids contamination from the great majority of Galactic stars. Of the data, objects that are detected in U_n are shown as circles with sizes proportional to brightness. The small triangles are those with limits only on $U_n - G$ colors (so that $U_n - G$ is an upper limit). About 20 Lyman break galaxies in the redshift range $2.7 \lesssim z \lesssim 3.4$ are selected from the data. The plot is courtesy of Chuck Steidel.

SEDs), and neural networks, based on training sets that automatically take into account Bayesian priors (Firth et al., 2003).

Photometric redshift determinations for faint galaxies can be split into two analysis-types; one fits the observed galaxy colors with redshifted template SEDs. The galaxy templates can be empirical, or stellar synthesis products, and (for large-z comparisons) can have an augmented Lyman region IGM opacity. The other technique by Connolly et al. (1995) is purely empirical using developed relations between galaxy colors, magnitudes and redshifts. Both methods have been quite successful in delivering prospective high-z candidate galaxies for a further critical spectral analysis using dispersed light. The artificial neural networks approach an r.m.s. redshift error of $\sigma_z \sim 0.02$ for redshifts $z \lesssim 0.35$. Recent comparisons of various methods can be found in Csabai et al. (2003) at small z ($z < 0.5$), and Firth

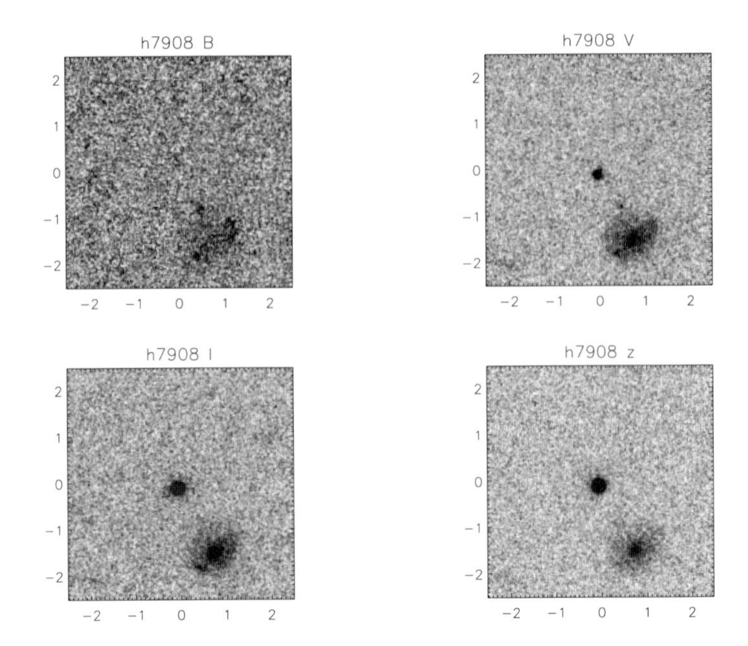

FIGURE 2: Central in each of these small area images we find a very compact, high surface-brightness distant galaxy ($z \simeq 5.19$) imaged in 4 HST photometric bands. Only a hint of the galaxy appears in B; it is substantially present in V, but much brighter in I and z. The diffuse galaxy below h7908 is a foreground galaxy, probably a spiral. The scale is in arc-seconds. Image courtesy of L. Moustakas.

et al. (2003) at larger z. Comparisons between most distant galaxy photo-zs, as they may be termed, and spectroscopic redshifts result in an accuracy of $\Delta z_{\rm rms} \approx 0.1(1+z)$ (see, for example, Stern & Spinrad, 1999, fig. 6). Firth et al., who use an artificial neural network and a training set, show errors as low as $\sim 5 \times (1+z)\%$ for simulated galaxies. For many purposes this sort of (even systematic) uncertainty is not a great flaw; of course at the largest photometric/spectroscopic redshifts the empirical comparison will be weakened as "training" magnitudes and colors relative to redshifts will be weakly constrained.

At yet larger redshifts, such as $z \simeq 5$, the faintness of the selected galaxy content leads to more foreground interlopers. This contamination results from a noisy SED because photometry, even with the HST, is imprecise at i and/or z-band magnitudes near 25 to 26. At this brightness level, most of the galaxies are still foreground to $z = 5$.

An extension of this procedure simply attempts to detect a faint IR (J, H, K, or a subset of them) galaxy image at a location with no optical signal (in R, i, z, for example). The likely interpretation would be either an extremely red galactic star, a dusty and heavily extincted galaxy (with red colors, then), or a Lyα discontinuity between the z-band (0.850 μm) and the J-band (1.2 μm), and thus a very large redshift between $z = 6.4$ and $z = 8.9$! Some as yet unconfirmed detections have

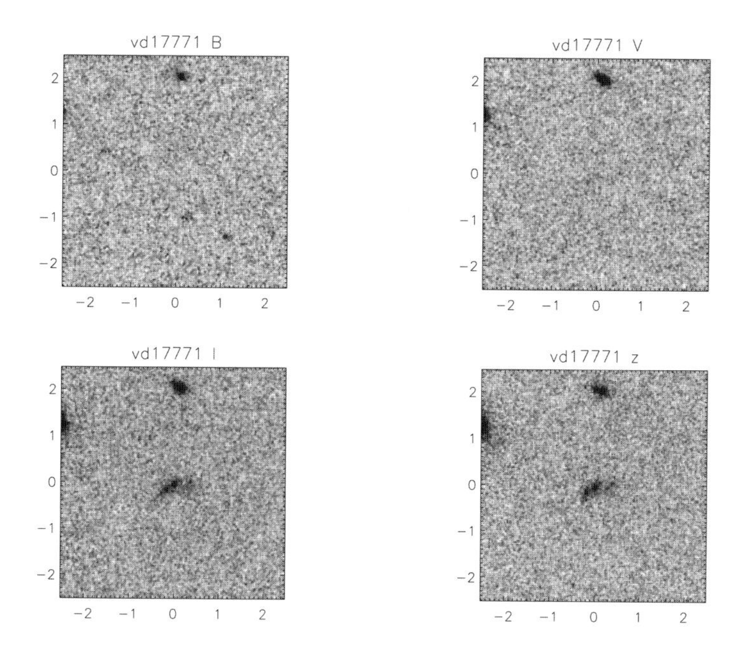

FIGURE 3: A structured galaxy imaged in B, V, I, z appears just-detectable in V, but is absent in B. It would be deemed a "V-drop". Almost no radiation escapes around the (rest frame) Lyman absorption series (observed V-band). This galaxy has the same redshift as that in Fig. 2. Scale is in arc-sec. Imaged with the ACS camera on HST, courtesy of L. Moustakas

resulted from the Hubble Ultra Deep Field (HUDF); they are constrained to lie in the range $7 \leq z \leq 8$ on the basis of their being undetected in z_{850} while being detected in the J-band (Bouwens et al., 2004). Any detection in the optical bands would likely rule out the very large redshift alternative. Photometric utilization of 3 IR filters can help to reduce the prevalence of foreground stars and red, but relatively nearby, galaxies.

Finally, Blain et al. (2003) studied the effects of dusty galaxies on photometric redshifts, finding that IR modeling offers almost no discrimination between the effects of dust and redshift. However, Sawicki (2002) suggests using the redshifted 1.6 μm H^- spectral feature as a photometric redshift indicator, with the Spitzer space telescope (previously termed SIRTF).

4.2.2 Using the Lyα Emission Line to Locate Extremely Distant Galaxies

Which distant galaxies show a strong Lyα emission line? In fact, a census formed from the U-drops ($z \sim 3$) of Steidel and collaborators (see also Shapley et al. 2003) indicate that large Lyα line luminosities are rather infrequent. Astronomers quantify the relative strength of the Lyα line by the equivalent width, which is

TABLE 2: GOODS[a] bandpasses and "drop"redshift ranges

Filter name	λ_{eff} (nm)	"drop" redshift
u	365	$\sim 3 \pm$
B	435	4 -4.5 $\pm$
V	606	5 $\pm$
i	775	~ 5.5
z	850	6+

a: Combines ground-based (u) and ACS filters.

given by the Lyα line luminosity divided by the continuum luminosity:

$$\mathcal{W}(z) = \frac{F_\alpha}{F_\lambda}(\text{Å}), \tag{2}$$

where F_α is in units of erg s^{-1} cm^{-2} and L_λ would then be in erg s^{-1} cm^{-2} Å^{-1}. Since F_α is redshifted away as $(1+z)^{-1}$, and F_λ redshifts as $(1+z)^{-2}$, the observed equivalent width is $(1+z)$ times the intrinsic equivalent width. Thus, if $\mathcal{W}_\lambda$ is the observed equivalent width, the intrinsic equivalent width (EW) is

$$\mathcal{W}_\lambda^0 = \frac{\mathcal{W}_\lambda}{(1+z)}. \tag{3}$$

Only ~11% of these continuum-selected galaxies around $z = 3$ have a Lyα line with a (rest-frame) $\mathcal{W}_\lambda^0 > 50$ Å (Shapley et al., 2003).

With about 1000 LBG spectra color-selected for $z \sim 3$, Shapley et al. (2003) find a strong inverse correlation between the reddening $E(B-V)$, and the EW of the Lyα emission line. The connection here is that Lyα is a resonance line, and multiple scatterings in a dust-laden H I cloud greatly increases the opportunity for absorption. Also interesting is the inverse correlation between the velocity offset of the Lyα emission line from the interstellar absorption lines and the EW of the Lyα emission line (e.g., Kunth et al., 1998). The velocity offset is characteristic of galactic winds spawned by supernovae.

The Shapley et al. (2003) results pertain to systems near or somewhat above L^*. However, as we shall see, the Lyα EWs of continuum-weak galaxies appears to be systematically larger than continuum-strong LBGs. By implication, a young, low metallicity galaxy with an as yet weak wind could be expected to produce high-EW Lyα emission. At present, the nature of the faint-end of the galaxy luminosity function at high redshift is poorly constrained. We will concentrate on these fainter "Lyα galaxies" in the next section.

Among the poorly documented manifestations of the trend of higher EWs with smaller galaxies is the surface density of Lyα emitters with large $\mathcal{W}_\lambda^0$ relative to the density of continuum-strong LBGs. Given the steepness of the faint-end slope parameter of the UVLF (Steidel et al., 1999) ($\alpha \simeq -1.6$ for both $z \simeq 3$ and $z \simeq 4$), one would expect the surface density of continuum-faint galaxies to be many times larger than standard LBGs ($R \lesssim 25.5$ mag). Yet the surface density of continuum weak, Lyα bright galaxies is of order the same as the surface density of LBGs (~1.5 or so per square arcminute per unit z in the magnitude range $21 \lesssim R \lesssim 25.5$ mag), suggesting that many continuum weak galaxies are also Lyα weak. Integrating the $z = 4$ LF, to a deep limit, one finds approximately 100 times more galaxies with continuum magnitudes in the range $25.5 \leq R \leq 28.5$, where the faintest would correspond to a SFR of about 0.5 to 1.0 $M_{\odot}\,yr^{-1}$.

Lyα emission-line detection can be done by using narrow-band filters, in conjunction with broad-band imaging, or by searches of archival spectra for serendipitous emission-line systems. Let us look briefly at these two techniques.

4.2.3 Narrow-band Detections

Narrow-band imaging enables one to estimate an equivalent width by making the approximation that the emission in the narrow band is dominated by an emission line, and that the broad band is dominated by the continuum luminosity. Often, the continuum is merely interpolated from two "straddling" broad-band fluxes. This technique has been used to detect emission-line galaxies, but experience has shown that follow-up spectroscopy is necessary to confirm the identification as Lyα emission. Figure 4 shows how windows in the OH airglow emission may be exploited with narrow-band filters to find very high redshift sources. Galaxies with lower redshifts and strong emission from forbidden lines can be major contaminants.

Deep narrow-band imaging has been quite useful in probing the faint-end of the Lyα emitters. For instance, Fynbo et al. (2003), who image Lyα in narrow bands at $z \sim 3$ to unusually low fluxes ($j_{\nu} \geq 7 \times 10^{-18}$ cgs) find many more lower-luminosity Lyα emitters than seen in the conventional LBG luminosity function. Follow-up spectroscopy has confirmed their redshifts, and shown that though their line luminosity is generally small, they have a large median (rest-frame) $\langle \mathcal{W}_{\lambda}^{0} \rangle \simeq 75$ Å, as shown in their fig. 7. They measure a large surface density ~10 galaxies per square arcminute per unit redshift. This is roughly a factor of 5 larger than the surface density of $z \sim 3$ LBGs to $R \sim 25.5$. These data suggest that the Lyα emission-line luminosity function is quite steep; reminiscent of the steepness of the UV LF at high redshift (Steidel et al., 1999). However, this conclusion is to be considered uncertain, as Hu et al. (2004) and the CADIS group, using narrow-band imaging to $z \simeq 5.7$ find much lower surface densities of emission-line galaxies, though it must be acknowledged that they probe the luminosity function less deeply than Fynbo et al. (2003).

If one integrates over the rest-frame UV LFs for redshifts $z \sim 3$ and 4 (Steidel et al., 1999) to the depth probed by Fynbo et al. (2003) (i.e., $R \sim 27$ mag) using the Steidel et al. steep faint-end slope ($\alpha = -1.6$), we find the predicted surface density of faint galaxies should be ~ 100 per square arcmin per unit z – roughly 50 times larger than the LBG surface density ($R \lesssim 25.5$) and ~10 times larger

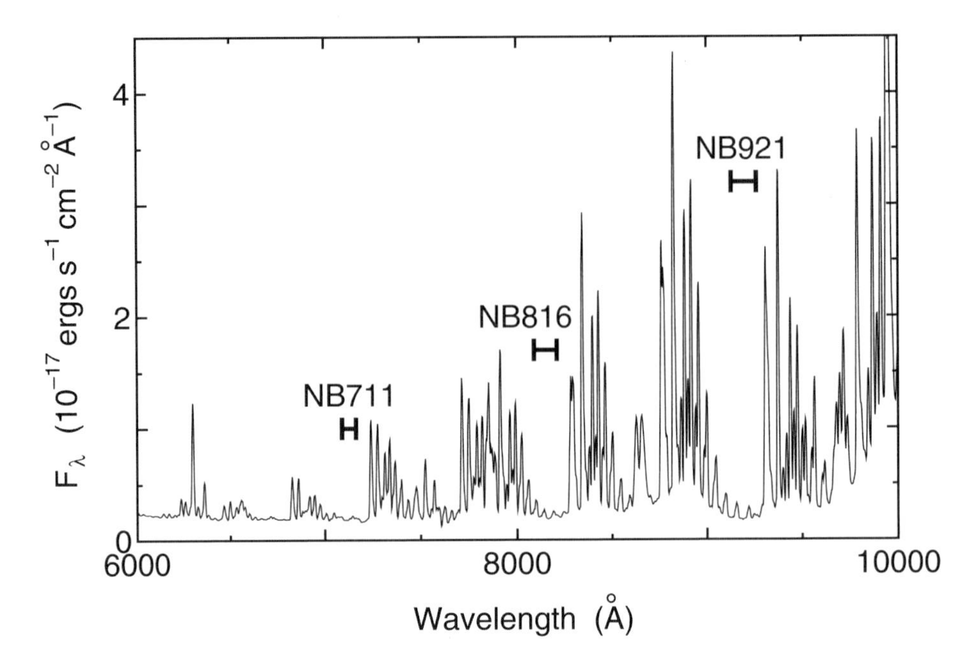

FIGURE 4: Optical spectrum of airglow emission, showing the placement of narrow-band filter transmission wavelengths employed at the Subaru telescope, and others, to take advantage of available windows in the airglow (note filter names, given in nanometers). The figure is from Taniguchi et al. (2003), reproduced by permission of the KAS.

than the Fynbo et al. (2003) findings for Lyα emitters. This result suggests either that only a relatively small fraction of continuum-weak galaxies may have large EW Lyα emission, or that the UV LF of Steidel et al. (1999) is excessively steep. We shall soon see (below), however, that Yan & Windhorst (2004) have estimated a faint-end slope of $z = 5.5$ to 6.5 galaxies to be even steeper.

In addition to these complexities, we must keep in mind that as we probe redshifts approaching $z = 6$, the objects existing then must have been maintaining the ionization of the Universe, and that those immediately before them must have been capable of re-ionizing it. In this connection, we note that Lehnert & Bremer (2003) find that $z \simeq 5$ LBGs do not supply enough UV radiation to keep the IGM ionized. Likewise, we learned above (Barger et al., 2003) that AGN cannot have re-ionized the Universe.

One enticing by-product of the Fynbo et al. result is that faint SF galaxies with strong Lyα may also emit considerable numbers of ionizing photons ($\lambda_0 < 912$ Å). If this hard radiation escapes the small, young galaxy, we may consider the likelihood that these numerous but faint SF galaxies could be responsible for the re-ionization of the IGM at redshifts just beyond our current "record z", near $z = 6.5$. Recent analysis from the Hubble Ultra Deep Field (UDF) (Stiavelli et al., 2004) suggests that they may have now detected i- and z'-drops, in the J and H NICMOS (a NIR HST camera) bands with a sufficient surface density to account for the re-ionization of the universe. There is some disagreement over the interpretation of this data (e.g., Bunker et al., 2004), so that more work must be done before the results of Stiavelli et al. are confirmed. Meanwhile, Yan & Windhorst (2004) find in excess of 100 redshift 5.5 to 6.5 candidates in the Hubble UDF, leading to a cumulative surface density of nearly 28 per square arcmin to an apparent magnitude of $z_{850} \sim 29.2$; corrections for incompleteness lead them to conclude that the faint-end of the LF has a slope $-1.8 \leq \alpha \leq -1.9$. If ionizing radiation escapes these objects, then their high surface density at that redshift implies they may be responsible for the re-ionization of the Universe.

A major uncertainty in the scenario for the ionization of the universe by faint sources is the requirement that a large fraction of the ionizing photons must escape from these small galaxies, for the clumping of hydrogen around the galaxy can intercept ionizing photons, and reprocess them into lower-energy photons. If we may ignore inhomogeneities, it is the recombination rate of this dense gas, proportional to the density of H II squared, that limits the reprocessing. If ionizing photons are emitted at a rate faster than the recombination rate, then the gas will become ionized, and the escape fraction of ionizing photons will increase. Observations of nearby starburst galaxies (Mas-Hesse et al., 2003) show that local velocity and density distributions of neutral gas (and dust), resulting from SN-driven outflows, can stochastically affect the reprocessing, and hence the escape fraction of UV photons from a galaxy. This problem presents a tough task for modelers of re-ionization. Recently, observations by GALaxy Evolution eXplorer (GALEX), a 0.5-m space telescope sensitive in the vacuum UV, found galaxies with characteristics quite similar to the Lyα galaxies seen at high-z (Heckman et al., 2005). The opportunity to study such low-z analogs may eventually solve the problem of the escape fraction of ionizing photons.

Another notable effort (Santos et al., 2004) systematically uses intervening rich clusters to find average magnifications of 10 or so, extending the depth of our reach to faint galaxies by an order of magnitude. Such a gain allows us to reach star-formation rates as low as 0.01 $M_{\odot}\,yr^{-1}$. This first assessment of the luminosity function of high-z Lyα lines finds that the Lyα LF is somewhat flatter than the halo mass function, implying a suppression of SF in low-mass halos.

With NB work there is a strong bias in favor of emission-line detections. NB measurements specifically look for the Lyα line, of course, but generally, these systems have very faint continua, and are unlikely to be positively detected without a prominent Lyα emission line. Color selection and the LBG technique does not select for emission lines, but requires a strong continuum for detection.

4.2.4 Distant Galaxies Detected as Lyα "Serendips"

Finally, we discuss a relatively passive method of location for very high-redshift galaxies. We spectroscopically sense the presence of a distant galaxy and measure its redshift by using small portions of the (galaxian) high-latitude sky as search areas for strong emission lines, including Lyα shifted into the yellow, red, and the near-infrared portion of the optical spectrum. The method has been successful, so far, at least out to $z = 5.2$, and with the Stern et al. (2005) Lyα galaxy, probably to $z = 6.545$. Of course we rely on a detection scheme adequate to address the accurate removal of (relatively) bright night sky emissions from our dispersed (typically 2-hour) exposures.

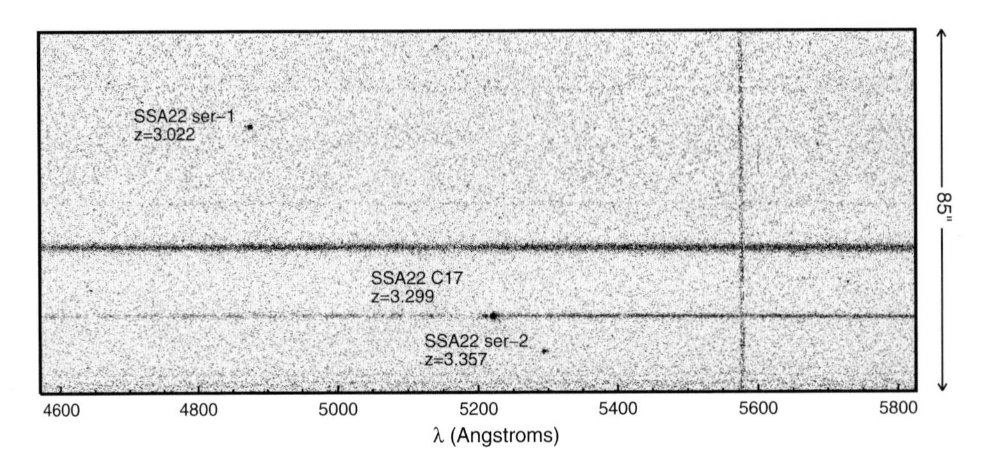

FIGURE 5: Serendipitously discovered Lyα galaxies in Keck LRIS spectra of the the galaxy C17 at $x = 3.299$ in the SSA2 field using a 600-line grating. The lack of an apparent continuum for either SER 1 or SER 2 indicates high-EW lines. The relatively high-resolution of the spectra enabled the [O II] doublet to be ruled out; Hα is ruled out because it would have to be blue-shifted – unprecedented for cosmologically placed sources. Image from Manning et al. (2000).

One such serendipitous sample from the HDF is discussed by Dawson et al. (2001). The galaxies are usually located on the basis of emission lines from H II regions. Lyα galaxies have been found in the range $2.442 \leq z \leq 5.77$ with a median redshift near $z = 3.9$. The relatively efficient spectral imaging of the Low Resolution Imaging Spectrometer (LRIS, Oke et al. 1995) on the Keck I and II 10-m telescopes, can detect emission lines with fluxes down to $\sim 10^{-17}$ erg s^{-1} cm^{-2}. With a 600-line grating ($R \equiv \lambda/\Delta\lambda \simeq 1200$) one can resolve the [O II] doublet and check the asymmetry found in many galaxy Lyα emission lines. That is important in ruling out common interlopers with [O II] or [O III] emission lines. The equivalent widths of the serendipitously discovered Lyα emission lines (we often refer to them as SERs) are quite large, since there often is no adequately detected continuum from the host system. Rest equivalent widths range from on order $\mathcal{W}_\lambda^0 \sim 20$ Å on the low contrast side, to ~ 350 Å (with large error bars) for the highest contrast emission lines.

Given the distribution of Lyα EWs found by Shapley et al. (2003), the Lyα line fluxes found in Dawson et al. (2001) would portend easily detectable continua. But their large equivalent widths argue here again for a differing parent population of line emitters compared to the normal continuum galaxies. A rather detailed view of what could be hiding at the high equivalent width and low continuum luminosity end of the Lyα galaxies can be found in Manning et al. (2000), in which two $z \gtrsim 3$ SERs are discovered in a single spectroscopic slit (refer to Fig. 5). Both Lyα lines have $\mathcal{W}^0_\lambda \geq 100$ Å (2-σ confidence level); one lacked any detectable line asymmetry. Neither have detectable continua; line fluxes are close to 1×10^{-17} erg s^{-1} cm^{-2}, implying SFRs of ~ 1 M$_\odot$ yr^{-1}. These objects are also physically quite small ($r \lesssim 500\ h_{50}^{-1}$ pc), consistent with the inside-out galaxy formation scenario (Cayón et al., 1996).

To further classify the SERs of $\mathcal{W}^0_\lambda \geq 50$ Å with a surface density, we have been locating about 2.3 ± 1 emission line galaxies per square arcminute per unit redshift at $z \approx 5$ (Dawson et al., 2001). If we could scale the number of anticipated LBGs on the basis of their $\mathcal{W}^0_\lambda$ distribution (Shapley et al., 2003) with $\mathcal{W}^0_\lambda \leq 40$ Å from our surface density of Lyα emission-line galaxies, we'd anticipate $\sim 20 \pm 10$ per square arc minute per unit z, far above the number actually measured in the normal two-color photometric manner. So, on the one hand, the surface density differences again seem consistent with a dwarfish faint continuum population with bright Lyα lines for the possible pre-galactic fragments (at $z \approx 5$). On the other hand, a "normal" population of more massive young stellar systems (the LBGs) with relatively weaker Lyα emission lines, appear to decline in number density as we look beyond $z = 5$ (see Bunker et al. 2003; Dickinson et al. 2004).

We conclude this section with a commentary on the observed/predicted Lyα line asymmetry, which pertains to serendips as well as targeted Lyα emitters. We have noted that the profiles of the Lyα emission lines in galaxies with detectable continua of all redshifts studied at high spectral resolution ($R = \lambda/\Delta\lambda \geq 1000$) are usually quite asymmetric. They show a long red tail or wing and a sharp cutoff on the violet spectral side (see Dawson et al. 2002; Rhoads et al. 2003). Interestingly one sees the same general profile locally, as is shown in Fig. 6 for IRAS0833+6517, $z \approx 0.02$. So the general Lyα forest cannot absorb all the "blame"; local gas associated with the galaxy plays a fundamental role in producing line asymmetry.

The asymmetry of the Lyα line can be measured in several ways. A simple and effective method is,

$$a_\lambda = \frac{\lambda_{10,\mathrm{r}} - \lambda_\mathrm{p}}{\lambda_\mathrm{p} - \lambda_{10,\mathrm{b}}}, \tag{4}$$

where λ_p is the wavelength of the line at its peak, $\lambda_{10,\mathrm{r}}$ is the wavelength on the red side at 10% of the peak flux, and $\lambda_{10,\mathrm{b}}$ is the wavelength on the blue side at 10% of the peak (Rhoads et al., 2003; Dawson et al., 2004). This parameter is almost always greater than 1 for Lyα lines, indicating an emission line with steeper blue-side, and a flatter red-side profile. The virtually ubiquitous asymmetry ($a_\lambda \geq 1.4$ or so) shown by the Lyα emission line has two uses: first as a confirmation (or perhaps a denial) of the line identification (Lyα is asymmetric, [O II] is a resolved doublet at medium spectral resolution, and [O III] 5007 is symmetric). A composite of 59 Keck/DEIMOS Lyα line profiles at $\langle z \rangle = 4.5$ is shown in Fig. 7, where the derived asymmetry index a_λ is shown according to Eq. (4). The mean rest-frame EW of component spectra is 58 Å. These emission-line spectra are a product of

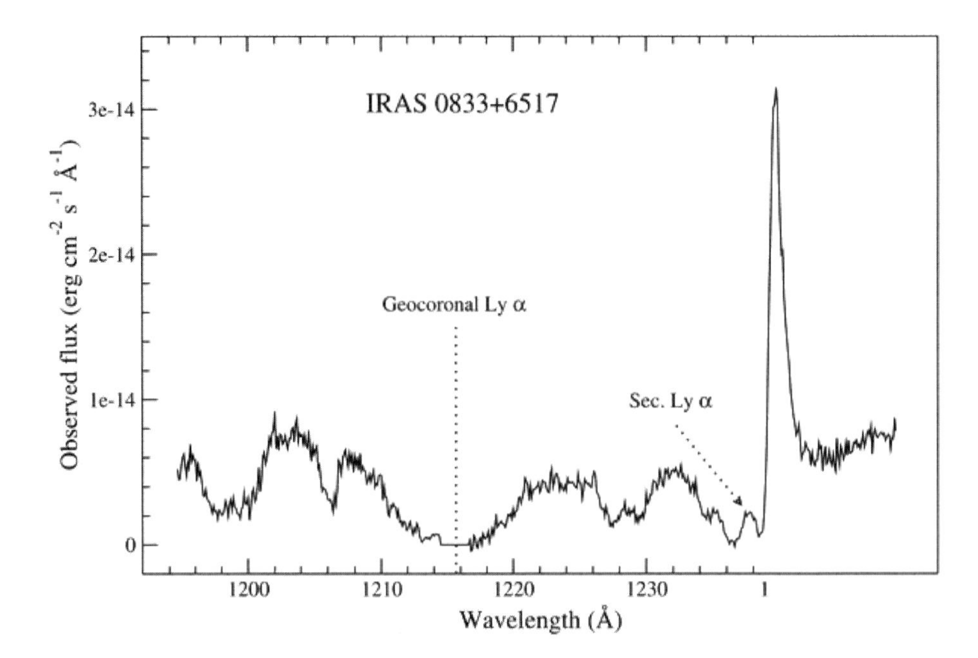

FIGURE 6: The spectrum from the central region of IRAS 0883+6517, a nearby galaxy at $z \sim 0.02$, showing a strong Lyα emission line, and a weaker (secondary) emission, perhaps caused by collisionally excited recombination, often found produced by winds in higher redshift Lyα galaxies. Note the asymmetry in the main emission line. Neutral hydrogen in the MW causes the broad geocoronal absorption line at $\lambda = 1216$ Å. Plot is courtesy of Mas-Hesse et al. (2003).

the Large Angle Lyα survey (LALA), a project led by S. Malhotra and J. Rhoads targeting certain low-extinction sight lines, such as the Lockman Hole.

Secondly, Lyα line-profiles are useful in guiding the development of theoretical work involving the expansion of neutral and ionized shells of gas accompanying SN- or SF-driven outflows in a young galaxy. The outflow is seen to evolve with time as does its Lyα spectral signature, according to spherically expanding models of the interstellar medium by Tenorio-Tagle et al. (1999), Ahn et al. (2002), Mas-Hesse et al. (2003), and Ahn (2004). So one might eventually be able to confidently model he "state of the starburst" from the strength and shape of the Lyα emission line.

However, we are concerned that the typical Lyα profiles are very alike over a span of objects selected to have a (nearly) common redshift, and between differing redshift ranges. One might anticipate the physical situation being strongly influenced by our viewing angle, or the specific time of the observation toward the star burst source (see Mas-Hesse et al. 2003) and thus create some variety of redshifted line-wings. Perhaps these will be observed as the resolution of observations and techniques for observing improve for $4 \leq z \leq 6$ galaxies with active ongoing SF. Of course, a theoretical escape from the embarrassment of the uniformity of line asymmetries might involve a long-lived spherical expansion (Tenorio-Tagle et al.

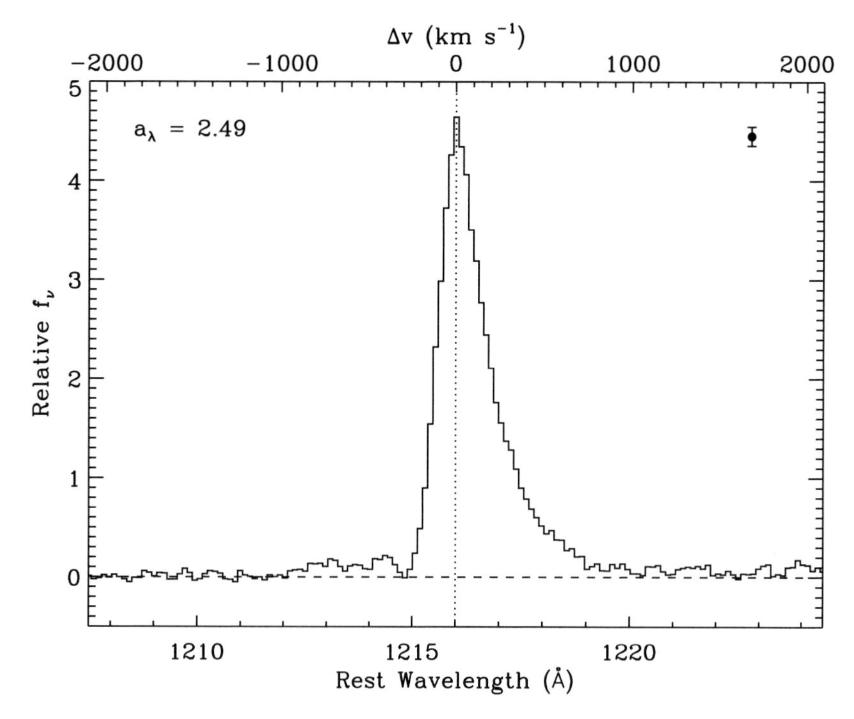

FIGURE 7: The composite spectrum of 59 confirmed Lyα emission lines ($\overline{z} \simeq 4.5$) from LALA galaxies (Dawson & Spinrad, in progress) observed with Keck/DEIMOS. The asymmetry is thought to be the result of outflowing neutral gas which scatters light from the red side of the Lyα emission line into the noticerable red wing at $\delta v \gtrsim 500$ km s^{-1}.. Plot courtesy of S. Dawson.

suggest the "correct phase" obtains ~1/3 of the time). However, that might give the appearance of an unpleasant interpretive "fine-tuning".

4.3 Record Redshifts

When observations are pushed to the most distant limits, the techniques for discovery are various; continuum breaks, narrow band, emission line (Lyα) or serendipity. In Table 3 we list published or otherwise secure "record redshifts" for galaxies. Most have prominent Lyα emission lines, this being the main method by which their redshift is confirmed. The first column lists the redshifts; the second lists the object designation; the third column lists the flux in units of 10^{-17} erg s^{-1} cm^{-2}. This is followed by the Lyα line luminosity in units of 10^{42} erg s^{-1}, calculated using a flat h $= \Omega_\Lambda = 0.7$ cosmology. The fifth column lists the discovery method: NB for narrow-band, LBG for color-selected Lyman breaks, and SER for serendipitously discovered systems. One system was discovered by slitless spectroscopy in the ~9100 Å airglow window. Since the publication of the review by Spinrad (2004), we note many new Lyα redshifts at $z > 5.5$, and our list now contains 30 objects, where the same interval earlier only had 18 detections. We do not expect this list to be exhaustive for long. Already there are about 4 z-drop candidates in the redshift range $7 \leq z \leq 8$ (Bouwens et al., 2004), and one from a cluster-magnified source at a probable $z \sim 7$ (Kneib et al., 2004). What is remarkable about the latter galaxy is its relatively strong SFR ($\sim 2.6\ \mathrm{M}_{\odot}\ \mathrm{yr}^{-1}$) juxtaposed with a very small angular diameter ($\lesssim 1.0 \mathrm{h}_{70}^{-1}$ kpc).

4.4 QSOs and AGN near the Contemporary Limits

Distant QSOs are of importance because they serve as bright beacons for understanding intervening gas clouds to large redshifts and because modern theory has the QSOs forming or growing a central BH of very large mass very early in their lives. SMBH might collapse from evolving isolated simulation peaks near $z \sim 10$ (Bromm, 2004). Let's look at the QSO detections approaching that (large) redshift.

The $z = 6$ "barrier" for QSOs fell with the application of the Sloan Digital Sky Survey long-λ (stellar) photometric indices and the following spectroscopy (Fan et al., 2001; Becker et al., 2001).

The i and z filters govern the range of temperature covered by cool galactic dwarf stars, and also the redshift of the Lyα line and the blue-ward Lyman forest absorptions with the extant Sloan photometric filters for i ($\lambda_{\mathrm{eff}} = 7615$ Å) and z ($\lambda_{\mathrm{eff}} = 9132$ Å). The redshift for the QSO, based on strong Lyα emission in the z-band and strong Lyα forest absorption in the i-band, yields a rough $z_{\mathrm{max}}(\mathrm{QSO}) = 6.5$; not surprisingly the observed QSO maximum z by this method is $z \sim 6.42$, for J1148+5251. To substantially surpass that large QSO redshift will require a different band pass and detector combination, probably in the infrared. For example, a $(J - z)$ color would be useful to $z_{\mathrm{Q}} \simeq 9$.

During the last few years there have been an increasing number of detections of molecular emission lines in distant QSOs. Of course this portends the discovery of rather massive reservoirs of gas (and probably dust) surrounding the QSO nucleus. Probably the host galaxy is also forming stars at this active epoch; however, differentiating SF from gravitational effects near an MBH will be difficult.

We might expect a small systematic shift of the molecular lines to a lower redshift with $\Delta z \simeq 0.01$ or 0.02 due to the scattering of the blue-ward side of the Lyα emission line. In the case of J1148+5251, the molecular line redshift, $z = 6.419$ is approximately as expected, relative to the $z = 6.43$ of the peak of the Lyα line. These more accurate or reliable CO redshifts may help us to decode the kinematics of neutral gas in proximity to the quasar.

Two other aspects of this result are of note. First, the presence of abundant CO molecules in this system suggests that these young systems have been quickly enriched in the "heavy" O, C, and maybe N elements – this at a time when the age of the universe is thought to be less than 900 Myr ($h = 0.7$). Another

TABLE 3: Census of Galaxies Confirmed at $z \gtrsim 5.5$

z	Source	f(Lyα)[a]	L(Lyα)[b]	Method[c]	Reference[d]
6.597	SDF J132522.3+273520	1.55	7.7	NB	1
6.585	SEXS1-SER	2.1	10.3	SER	2
6.580	SDF J132432.5+271647	0.62	3.07	NB	1
6.578	SDF J132418.3+271455	2.1	10.4	NB	3
6.578	SDF J132518.8+273043	0.73	3.61	NB	1
6.56	HCM 6A	0.65[c]	3.2	NB	4
6.554	SDF J132408.3+271543	1.66	8.13	NB	1
6.542	SDF J132352.7+271622	0.73	3.56	NB	1
6.541	SDF J132415.7+273058	1.53	7.49	NB	1,2[e]
6.540	SDF J132353.1+271631	0.82	4.00	NB	5
6.535	LALA J142442.24+353400.2	2.3	11	NB	6
6.518	KCS 1166	1.9	9.2	slitless	7
6.506	SDF J132418.4+273345	1.94	9.35	NB	1
6.33	SDF J132440.6+273607	4.0	18.2	LBG	8
6.17	LAE@ 0226-04 Field	3.9	16.7	LBG	9
5.87	BDF1:19	0.31	1.2	LBG	10
5.83	CDFS 5144	1.6	6.0	LBG	11
5.79	GLARE 3001	0.76	2.8	LBG	12
5.783	CDFS SBM03#3	2.0	7.33	LBG	13
5.746	LALA5 1-03	1.9	6.7	NB	6
5.744	BDF1:10	2.4	8.7	LBG	15
5.74	SSA22-HCMI	1.7	6.1	NB	14
5.700	LALA5 1-06	3.9	13.8	NB	6
5.687	LAE J1044-0130	1.5	5.2	NB	15
5.674	LALA5 1-5	2.7	9.3	NB	6
5.655	LAE J1044-0123	4.1	14.3	NB	5
5.649	BDF2:19	2.5	8.7	LBG	10
5.631	F36246-1511	2.3[f]	8.0	SER	16
5.60	HDF 4-473	1.0	3.4	LBG	17
5.576	Abell 2218 lens	0.85[g]	0.6	LBG	18

[a] Lyα flux in units of 10^{-17} erg s^{-1} cm^{-2}.
[b] Lyα line luminosity in units of 10^{42} erg s^{-1}.
[c] Discovery method: NB – narrow-band, LBG – Lyman break, SER – serendipity.
[d] 1. Taniguchi et al. (2005), 2. Stern et al. (2005), 3. Kodaira et al. (2003), 4. Hu et al. (2002), 5. Taniguchi et al. (2003), 6. Rhoads et al. (2003), 7. Kurk et al. (2004), 8. Nagao et al. (2004), 9. Cuby et al. (2003), 10. Lehnert & Bremer (2003), 11. Dickinson et al. (2004), 12. Stanway et al. (2004), 13. Bunker et al. (2003), 14. Hu et al. (1999), 15. Ajiki et al. (2004), 16. Dawson et al. (2001), 17. Weymann et al. (1998), 18. Ellis et al. (2001).
[e] The flux and luminosity is the average of measurements by authors 2, the discoverers, and 1.
[f] Flux via personal communication with S. Dawson, 2004.
[g] Two lensed systems of nearly the same flux. We show here the average intrinsic flux, with amplification factor ~32 removed.

interesting aspect is the QSOs' extremely high luminosity, with $M_{1450} \sim -27.6$. If it is not gravitationally lensed, the luminosity would suggest a BH mass of around several billion solar masses, and a dark matter halo of 10^{13} M$_\odot$ (Fan et al., 2003). High-resolution images of this and other $z > 6$ QSOs (with seeing better than

$0''.5$) found only single images; there were no signs of multiple (lensed) images. However, the presence of strong C IV absorption at $z = 4.95$ suggests it could be at least weakly magnified. However, the three $z > 6$ SDSS QSOs are all very bright, and it is improbable that all have been similarly magnified. Achieving such a large over-density at such an early time must be theoretically reconciled with the UDF observations of physically small and faint z-drops that are thought to have stimulated the re-ionization of the universe (Stiavelli et al., 2004). This issue may pose a significant problem for theorists.

In any case, arguably the most important use of the $z > 6$ QSOs is to study distant intervening gas – primarily the H I IGM clouds of the various Lyman bound-free forests (just below Lyα, Lyβ, Lyγ toward the ultraviolet at rest wavelengths). The point here is that the Lyman edges of the IGM absorptions are quite strong, perhaps saturated across sizable wavelength (and thus, redshift) intervals. A key question is, has the IGM at $z > 6$ been fractionally a neutral gas at these early epochs?

In reverse of z-order, the time history of the "dark ages" and the eventual formation of stars and re-ionization, leads us to well-accepted Universal models of the IGM. We envision zero-metallicity luminous hot stars (called 'Population III') causing the ionization over discrete "nests" in the growing structures of matter at z from $\sim$30 to $z \leq 8$. This is seen in contemporary galaxy growth simulations. Though theoretically all seems in order, observationally, the epoch of ionization appears to be more locally varied.

The heavy absorption, low-transmission regions blue-ward of the Lyα line in the high redshift QSOs remind us of the Gunn–Peterson projections (Gunn & Peterson, 1965) for neutral H gas absorption in the IGM. Portions of high-redshift QSO spectra (i.e., the "Sloan" QSOs at $z = 6.28$, 6.39 and 6.42) imply a partially neutral state at redshifts slightly below the emission-line redshift (see Fig. 8). Because QSOs are a strong source of ionizing radiation, the ionization state in the immediate proximity of the QSO is not representative of the IGM. This is called the proximity effect. Thus, a region below $\lambda \approx 1100$ Å in the QSO restframe is the best place to search for a true Gunn–Peterson "neutral path", a region far enough from the target QSO to not yet have been re-ionized.

The searches for "dark spectral patches over Δz" in the Gunn–Peterson key epochs of near-neutrality in the IGM has been taken up by Djorgovski et al. (2001); White et al. (2003), and Songaila (2004). The drop in emission from the intrinsic continuum, projected from the long λ-side of the Lyα emission line to below it, can be used to calculate the average optical depth for Lyα and Lyβ radiation at $z \gtrsim 6$. Both White et al. (2003) and Songaila (2004) find Lyα optical depths $\tau \gtrsim 6$ for $z > 6$, implying transmission of only about 1/4 of 1% of blue-ward continuum photons redshifted to $\lambda \approx 1216$ Å in the IGM. Figure 9 shows the trend of Lyα optical depths taken from Sloan QSOs. The figure shows a pronounced departure (beginning at $z \sim 5.7$) from the trend of lower-redshift measurements (solid line).

We end this section with a speculative flourish. It would be challenging and important to locate QSOs at a larger redshift ($z > 6.4$), well enmeshed in the diffuse H I network, and measure its span of lowest transmission $\Delta\lambda$ regions.

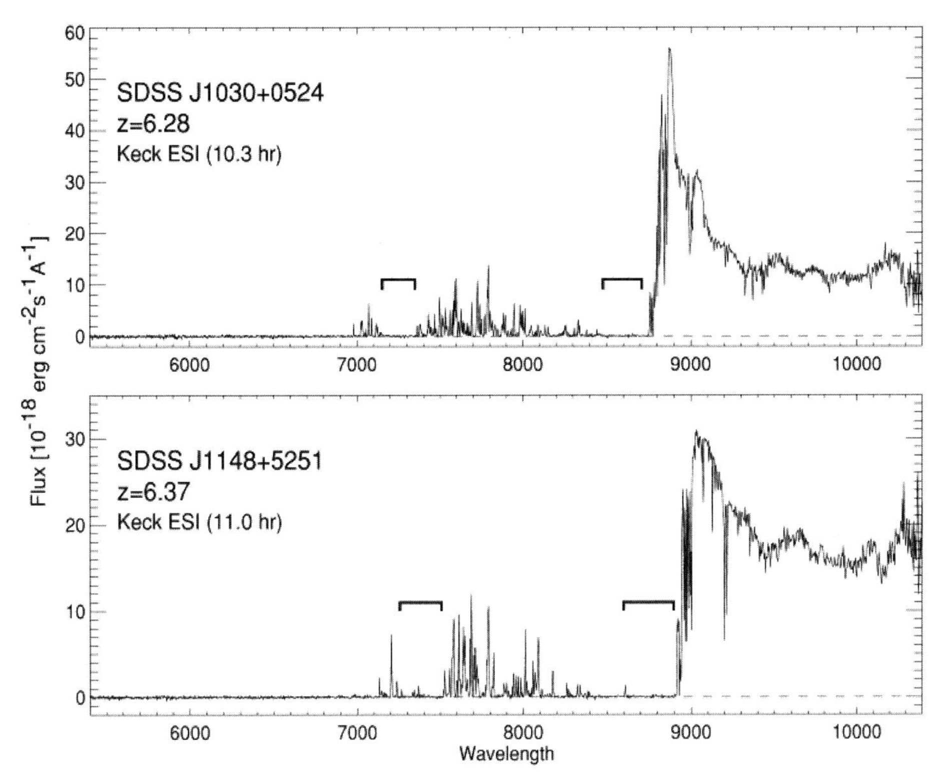

FIGURE 8: The spectra of two SDSS QSOs with significant Gunn–Peterson troughs near their Lyα and Lyβ lines, indicating a substantially neutral IGM zone. Plot is from White et al. (2003).

The exact manner and redshift of re-ionization is ambiguous since it deals with a very limited sampling of high-redshift space and a definitional uncertainty of what exact level of re-ionization would satisfy our criteria for the onset of an ionized Universe. One *could* say that we have already arrived at this epoch; the HUDF probes deeply into the LF of star-forming galaxies, finding a few sources which suggest that we may have found the main sources of re-ionization (Bouwens et al., 2004). Further, our probes of that re-ionization – the few QSOs at $z \gtrsim 6$ – are showing a rapid transition from a generally un-ionized IGM to a highly ionized IGM at $z < 6$. Perhaps, however, a more cautious statement is in order. We are probing *an* epoch of re-ionization; as Cen (2003) suggests, an earlier round of Population III stars in young galaxies with a top-heavy IMF may have effectively re-ionized the universe. In his scenario, re-ionization began at $z \approx 30$, and was completed by $z \sim 15$. Following this perhaps self-limiting burst, it would have become largely neutral again. This scenario has the benefit of explaining the recent Wilkinson Microwave Anisotropy Project (WMAP) polarization data that suggests re-ionization at $z \sim 15$. The top-heavy IMF could produce metal-free Population III stars of mass 100–250 $M_{\odot}$. The burst in early SF is expected to

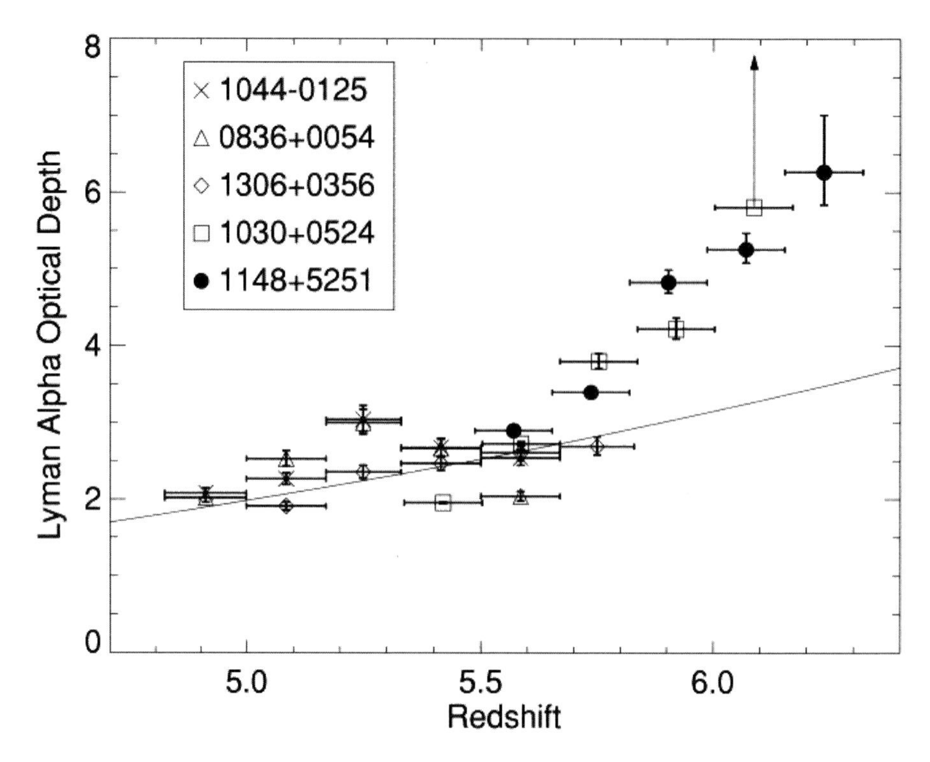

FIGURE 9: The Lyα optical depth as a function of redshift from a sample of high-redshift QSOs. The solid line represents the evolution based on a Lyα cloud line density proportional to $(1+z)^{5/2}$. The quasar J1030+0524 systemic redshift is $z = 6.28$, and J1148+5251 is at $z = 6.418$. Plot is from White et al. (2003).

result in clusters of BHs in the range of 100–200 $M_\odot$– seeds, perhaps, for the giant MBHs thought to be present in the Sloan $z > 6$ QSOs.

4.5 Galaxy Morphology at High-z

We have discussed (Chapter 3) attempts to discern galaxy morphology at high redshift by systematic means with the CAS system. However, the current robustness of the technique at very high redshifts leaves much to be desired. Nevertheless, our investigations of high-z galaxies afford us some general conclusions. As we probe to higher redshifts, particularly with the optical searches, we note an increasing irregularity of the galaxies detected. Some of this may have to do with a likely increasing merger rate with redshift, but we must be aware that the wavelengths we are sampling in the optical originate from farther and farther into the blue and UV as we look back. Local galaxies, when viewed in IR light have a smooth appearance since this light traces the older stellar populations, which have had the opportunity to "migrate". On the other hand, the UV-bright portions of local spiral galaxies can be quite blotchy and almost random-looking. We must carefully parse these two aspects of the irregularity of galaxies as we interpret high redshift objects. In general, however, we can say that the grand-design spirals are almost never detected with confidence at $z > 2.5$ and LBGs may dwindle in fractional representation as redshifts approach $z = 6$. What we see in increasing numbers as we probe deeper redshifts are small Lyα galaxies, which have diameters of about 1 kpc. We have called these compact first generation galaxies (Chapter 3, §3.4). In the context of a model of hierarchical assembly, they can be considered as "seeds", which may constitute the "building blocks" from which the LBGs and grand-design spirals to come are built.

4.6 Large scale Structure at High-z

The large-scale structure of luminous systems at high redshift is difficult to extract from a Universe seen from the myopic perspective of our near-Earth telescopes. At low-redshift, the LSS appears to consist of massive condensations that are relatively evenly distributed, with an interconnecting "web" of filaments, and even "walls", which leave a complementary structure of filaments and voids of size approaching 100 Mpc diameter. However, at high redshift, with spectral or photometric redshifts, we are just able to tease the underlying structure from the confusion. The clusters, seen to $z \lesssim 1$, are not seen at very high redshift, except as proto-clusters. Proto-clusters can be detected in redshift surveys, or by the enhanced surface densities of Lyα galaxies (Steidel et al., 1998; Venemans et al., 2002; Shimasaku et al., 2003). At $z \sim 4$, Venemans et al. find a concentrated protocluster with ~ 20 Lyα galaxies within a radius of 1.3 Mpc and an over-density of Lyα galaxies of ~ 15. Shimasaku et al. find an over-density of Lyα galaxies of ~ 2 at $z \simeq 5$. While the main concentration is spread over ~ 12 Mpc, a lower-density "filament" of Lyα galaxies extends ~ 20 Mpc. Of 43 candidates, six were spectroscopically observed, and five were confirmed, suggesting a low contamination rate among the detections.

That the surface density of Lyα galaxies has a large cosmic variance is suggestive of significant clustering on large scales (some megaparsecs). In Manning et al. (2000), it was suggested that when clusters begin their early stages of collapse, the increased pressure may trigger the collapse of "critical" Bonnor–Ebert (isothermal) spheres (if they exist), causing concentrated SF, as noted in §3.2. If this model is correct, then there may exist many sub-galactic clumps contemporary to the proto-clusters that lie between the mass concentrations, and so do not have the conditions to stimulate the collapse and subsequent SF, but may do so later as the Universe evolves.

4.7 Summary

The task of discriminating the high redshift galaxies (with $z \gtrsim 5$) involves a range of techniques, which we reviewed in this chapter.

In the radio wavelengths , we find that steep radio synchrotron spectral indices are associated with high redshift radio-loud AGN. However, their comoving number density is found to decline strongly with redshift, and very few new high-z detections are expected.

In the most energetic of wave-bands, gamma ray bursts are thought to occur in star-forming galaxies, and are likely associated with SNe. As the astronomical network of observatories that are capable of quick response to space-based detections of GRBs becomes perfected, we will learn more about these elusive beacons.

The X-ray band is useful for the study of AGN. The high spatial resolution of Chandra has enabled the discovery of a population of $z \gtrsim 4$ AGN associated with faint galaxies, but at a rate of only $1/50^{\text{th}}$ that of expectations, based on the AGN LF.

In the millimeter and sub-millimeter regions, SCUBA, and other planned detectors can take advantage of large negative K-corrections on the Rayleigh–Jeans side of the dust-emission spectrum to have a relatively flat selection function in z. SCUBA detections are likely to be associated with LIRGS or ULIRGS (ultraluminous IR galaxies), often with no optical identification. The current low spatial resolution results in source confusion, a problem to be solved by detectors and telescopes now under planning or construction.

Traditional optical searches include the Lyman break dropout, detection of Lyα emission lines, and optical identification of AGN. The dropout technique is the principal selection criterion for high-redshift objects, since intergalactic Lyα clouds at high redshift are very effective in removing Lyman continuum radiation from the QSO beam, leaving a "break" that is indicative of the redshift of the object. The same result occurs just short of the Lyα line. Detecting this break at increasing redshifts requires chasing the observed-frame break to ever longer wavelengths, a task requiring new technology and techniques as one encounters the Earth's interfering atmospheric lines (OH emission, mainly) and reduced detector quantum efficiency.

The Lyα emitting galaxies are star-forming galaxies that are quite often in their nascent stages. Often, a lone line will be seen in spectra, with a weak or undetectable continuum. Yet because it is a resonance line, it may be scattered by the surrounding H and absorbed by the neutral gas and dust. The scattering may be on physical scales of the nucleus, a galactic wind or the IGM – or all three. The result is normally an asymmetric emission line, itself a useful diagnostic of the galaxy's gas kinematics. Serendipitous searches provided intriguing information about the source population, prompting the development of a new technique designed to seek out these galaxies. For instance, Egami et al. (2005) use the magnifying power of the rich cluster A 2218 to detect the weak continuum a very small galaxy ($M_{\text{stars}} \sim 10^9$ M$_\odot$) with no detected Lyα in emission at $z \sim 6.7$. Coordinated narrow-band imaging in conjunction with broad-band imaging can provide a list of possible Lyα galaxies within a small redshift range, to be confirmed by subsequent spectroscopy.

Optical QSOs at very high redshift have been detected with the Sloan survey. These few QSOs have provided invaluable information on the ionization state of the IGM at the rough redshift of re-ionization. New techniques for the calculation of the Lyα optical depth suggest that the epoch of re-ionization may have been at about $z = 6.5$. However, pockets of Strömgren spheres at higher redshift may exist, leaving the hope of detection of Lyα galaxies at still higher redshift, hopefully beyond $z = 7$.

REFERENCES

Ahn, S. 2004, ApJL, 601, L25

Ahn, S., et al., 2002, ApJ, 567, 922

Ajiki, M., et al., 2004, PASJ, 56, 597

Andersen, M. I., et al., 2000, AAp, 364, L54

Barger, A. J., et al., 2002, AJ, 124, 1839

Barger, A. J., et al., 2003, ApJL, 584, L61

Barger, A. J., et al., 2003b, AJ, 126, 632

Baum, W. A. 1962, in *Problems of Extragalactic Research*, ed. G. C. McVittie, Vol. 51 (New York: Macmillan, IAU Symposia), 390

Becker, R. H., et al., 2001, AJ, 122, 2850

Blain, A. W., et al., 2003, MNRAS, 338, 733

Bloom, J. S. 2003, AJ, 125, 2865

Bouwens, R. J., et al., 2004, ApJL, 616, L79

Bremer, M. N., et al., 2004, MNRAS, 347, L7

Bromm, V. 2004, PASP, 116, 103

Bunker, A. J., et al., 2004, MNRAS, 450

Bunker, A. J., et al., 2003, MNRAS, 342, L47

Cayón, L., et al., 1996, ApJL, 467, L53

Cen, R. 2003, ApJ, 591, 12

Chambers, K. C., et al., 1990, ApJ, 363, 21

Chapman, S. C., et al., 2003, Nature, 422, 695

Chapman, S. C., et al., 2005, ApJ, 622, 772

Connolly, A. J., et al., 1995, AJ, 110, 2655

Conselice, C. J., et al., 2003, ApJL, 596, L5

Csabai, I., et al., 2003, AJ, 125, 580

Cuby, J.-G., et al., 2003, AAp, 405, L19

Dawson, S., et al., 2004, ApJ, 617, 707

Dawson, S., et al., 2002, ApJ, 570, 92

Dawson, S., et al., 2001, AJ, 122, 598

De Breuck, C., et al., 2000, Astron. Ap., 362, 519

De Breuck, C., et al., 1999, Astron. Ap., 352, L51

Dickinson, M., et al., 2004, ApJL, 600, L99

Djorgovski, S. G., et al., 2003, ApJL, 591, L13

Djorgovski, S. G., et al., 2001, ApJL, 560, L5

Egami, E., et al., 2005, ApJL, 618, L5

Ellis, R., et al., 2001, ApJL, 560, L119

Fan, X., et al., 2003, AJ, 125, 1649

Fan, X., et al., 2001, AJ, 121, 54

Firth, A. E., et al., 2003, MNRAS, 339, 1195

Fynbo, J. P. U., et al., 2003, A & A, 407, 147

Giavalisco, M., et al., 2004, ApJL, 600, L93

Gunn, J. E. & Peterson, B. A. 1965, ApJ, 142, 1633

Haiman, Z. & Loeb, A. 1998, ApJ, 503, 505

Heckman, T. M., et al., 2005, ApJL, 619, L35

Hu, E. M., et al., 2004, AJ, 127, 563

Hu, E. M., et al., 2002, ApJL, 568, L75

Hu, E. M., et al., 1999, ApJL, 522, L9

Kneib, J., et al., 2004, ApJ, 607, 697

Kodaira, K., et al., 2003, PASJ, 55, L17

Koo, D. C. 1985, AJ, 90, 418

Kreysa, E., et al., 1998, in Proc. SPIE Vol. 3357, p. 319-325, *Advanced Technology MMW, Radio, and Terahertz Telescopes*, Thomas G. Phillips; Ed., 319–325

Kunth, D., et al., 1998, AAp, 334, 11

Kurk, J. D., et al., 2004, AAp, 422, L13

Lehnert, M. D. & Bremer, M. 2003, ApJ, 593, 630

Manning, C., et al., 2000, ApJ, 537, 65

Mas-Hesse, J. M., et al., 2003, ApJ, 598, 858

McCarthy, P. J. 1999, in *The Most Disant Radio Galaxies*, eds. H. Röttgering, P. N. Best & M. D. Lehnert (Dordrecht: Kluwer)

Nagao, T., et al., 2004, ApJL, 613, L9

Oke, J. B. et al. 1995, PASP, 107, 375

Omont, A., et al., 2003, AAp, 398, 857

Rhoads, J. E., et al., 2003, AJ, 125, 1006

Santos, M. R., et al., 2004, ApJ, 606, 683

Sawicki, M. 2002, AJ, 124, 3050

Shapley, A. E., Steidel, C. C., Pettini, M., & Adelberger, K. L. 2003, ApJ, 588, 65

Shimasaku, K., et al., 2003, ApJL, 586, L111

Smail, I., Chapman, S., Blain, A., & Ivison, R. 1991, in *Maps of the Cosmos*, ed. M. Colless & L. Staveley-Smith, Vol. 00 (San Francisco: ASP Conference Series), 11

Songaila, A. 2004, AJ, 127, 2598

Spinrad, H., 2004 in *Astrophysics Update*, ed. J. W. Mason, (Heidelberg:Springer-Praxis), p. 155

Spinrad, H., et al., 1995, ApJL, 438, L51

Stanway, E. R., et al., 2004, ApJL, 604, L13

Steidel, C. C., et al., 1998, ApJ, 492, 428

Steidel, C. C., et al., 1999, ApJ, 519, 1

Steidel, C. C., et al., 2003, ApJ, 592, 728

Steidel, C. C., et al., 1996, AJ, 112, 352

Steidel, C. C. & Hamilton, D. 1992, AJ, 104, 941

Stern, D., et al., 1999, AJ, 117, 1122

Stern, D. & Spinrad, H. 1999, PASP, 111, 1475

Stern, D., et al., 2005, ApJ, 619, 12

Stiavelli, M., et al., 2004, ApJL, 610, L1

Taniguchi, Y., et al., 2003, JKAS, 36, 283

Taniguchi, Y., & 39 co-authors 2005, PASJ, 57, 165

Tenorio-Tagle, G., et al., 1999, MNRAS, 309, 332

van Breugel, W., et al., 1999, ApJL, 518, L61

Venemans, B. P., et al., 2002, ApJL, 569, L11

Waddington, I., et al., 1999, ApJL, 526, L77

Weymann, R. J., et al., 1998, ApJL, 505, L95

White, R. L., et al., 2003, AJ, 126, 1.

Williams, R. E., et al., 1996, AJ , 112, 1335

Yan, H. & Windhorst, R. A. 2004, ApJL, 612, L93

Chapter 5

Observational Cosmology

Our studies of the Universe on large scales have led us to an unusually robust set of observations dealing with the astrophysics of the expansion as seen in the "kinematics" of the redshift, with supernovae as "standard candles", and in the interpretational details of the fluctuation scales superposed on the 2.7 K cosmic microwave background (CMB).

In this chapter we discuss the Hubble (M, z) diagram using the type 1a supernovae, and the surprising conclusion that a force resembling a cosmological constant plays a dominant role in the late-time expansion of the Universe. Also we consider the cosmological density of baryons, and the density of its nucleonic mix, especially the light elements (^{4}He, ^{3}He, ^{2}H = D) in the IGM and in galaxy H II regions. In addition, we compare the expansion age of the Universe and the oldest stellar ages. The expansion age has a complex dependence on the assumed cosmology as well as an inverse relation to the Hubble constant, while the age of the oldest stars comes from our improving knowledge of stellar evolution, the distance scale in the Milky Way, and nuclear reaction rates. Such a comparison provides an important check on the internal self-consistency of our understanding of the Universe and its evolution. Many of these parameters have recently been estimated by the analysis of the CMB fluctuations, and the most recent results will serve as a centerpiece for our discussion.

Observational cosmology is a field in which observers and cosmologists try to make contact, and produce a "concordance" model of the Universe. It will be useful to review the cosmological scenario known as the Standard Model. The Standard Model, however, is not one thing or a set of parameter values, but the evolving consensus of astronomers.

We start with perhaps the simplest cosmological model that features expansion, the Einstein–de Sitter model. The expansion law for this model (i.e., the Friedmann equation) is,

$$H = \left(\frac{\dot{a}}{a}\right) = \sqrt{\frac{8\pi G}{3}\rho_{\rm crit}}, \tag{1}$$

where H is the expansion parameter (or Hubble constant), a is the proper physical scale factor of the Universe, and $\rho_{\rm crit}$ is the critical matter density (actually, $\rho_{\rm crit}$ must also include the mass-equivalent of the energy density as well). This is the expansion law of the Einstein–de Sitter model. For a matter-dominated Universe, it is thought to be a very good approximation in the early Universe. What makes the Einstein–de Sitter a critical density Universe is that it lies in between two antithetical cosmology types. One, with a density lower than $\rho_{\rm crit}$, has a negative curvature (it is "open", or "hyperbolic"), and one with a supercritical density has a positive curvature ("spherical"). The curvature for the Einstein–de Sitter model is zero, and cannot change within the constraints of the model. Conversely, once the curvature is positive, it can never become negative within the constraints of the Friedmann model. A zero-curvature model is spatially infinite.

But the problem which caused the eventual abandonment of the Einstein–de Sitter Universe is not its infinite volume, but subtle inconsistencies. One is the "flatness" problem, and another is the "horizon" problem; both are sometimes subsumed into the so-called "fine-tuning" problem. The flatness problem is invoked by considering the instability of a homogeneous critical density in an evolving Universe. Because the Universe is thought to come from a very dense state, the Universe would have had to be fantastically uniform near the beginning in order to avoid massive gravitational collapses. If we are anywhere near a critical density today, then at the time of decoupling density variations could only be one part in 10^{16}, or one part in 10^{60} at the Planck epoch. Either it is an extreme coincidence that the present density is close to one, or there must be a mechanism to explain its flatness. The horizon problem* refers to the unlikelihood of the isotropy of the CMB far beyond the causal horizon at the epoch of decoupling in a standard Big Bang universe. Though the CMB is very isotropic, on angular scales of only about 1°, regions are causally distinct during the epoch when the CMB radiation decoupled from matter, under a Big Bang theory. Therefore the observed uniformity of the CMB could not have evolved naturally by thermalization within the constraints of the Einstein–de Sitter model.

To solve these problems, a non-singular model of creation, dubbed *inflation*, was grafted onto the Big Bang theory (for instance, Guth, 1981). In the inflation model, space is caused to expand in a roughly exponential way. If this is done in a way that is in conformance with Eq.(1), then deviations from the mean density, which would otherwise have caused primordial black holes to form, are driven exponentially toward zero, resulting in the flatness and horizon problems being solved. The former was just explained, while the latter is explained by realizing that the currently observable Universe should be the inflated remnants of a very small piece of the early inflating Universe, and hence the causal connection is straightforward.

The rationalization of the physical process that results in the inflationary field is based on abstract physical theories that evoke a "super-symmetry" on

*The horizon is the limit of the distance at which phenomena can be causally effective. A good rule of thumb for the optical horizon size is $R_H = c/H$, where c is the speed of light and H is the expansion parameter. Today, $R_{\rm H} \simeq 4000\ h^{-1}$ Mpc.

the quantum-mechanical level. This physical theory is beyond the ken of simple extragalactic astronomers. We present a simplified version: all we really need to know is a few simple rules for inflation: first, it must obey the energy density form of Eq. (1), and second, it must be in the positive direction (i.e., it is creative). If an inflation of the Universe began when the energy density ($E_0 = \rho_{\rm crit}\, c^2$) was very was very large, then the expansion parameter was large also, $H = \sqrt{8\pi G E_0/3c^2}$.

As it expands, energy must be added, for with the large expansion rate, the energy density driving the high inflation rate would otherwise be diluted, and the expansion would revert to the "coasting" style of the Big Bang model, according to the strictures of Eq. 1. On the surface, it looks as though there were a violation of the principle of conservation of energy. Somehow, the increasingly negative potential energies between each pair of particles in the Universe, or an energy for space itself, must be invoked to balance the energy equation.

While some cosmologists think that energy conservation is very important, others appear to be comfortable with dropping it when considering the Universe as as a whole. We leave these speculations for philosophers.

Supersymmetry theories envision a unification of fields at a GUT ("Grand Unified Theory") temperature of $T \approx 10^{28}$ K. Above this temperature, all the fields of physics are thought to be unified, while at a lower temperature, there is a symmetry breaking which introduces the phase change that inflation theorists invoke to explain inflation. The energy densities consistent with such a high temperature imply a huge inflationary expansion parameter. The details, and the mode of its initiation and conclusion, are beyond the scope of this book.

Guth's inflation model did well in explaining the main problems with the Big Bang theory that were known at that time. However, the original inflationary model was difficult to reconcile with observations of large-scale structure in the low-redshift Universe. Inflation had done its job of flattening inhomogeneities in the early Universe all too well. As early simulations showed, it was difficult to produce the galaxy formation seen at high redshift, or the mature galaxies, clusters and large-scale structure seen locally, with the post-inflation Universe so perfectly flat. Thus, the horizon and flatness problems were replaced by a large-scale structure problem.

New inflation (Guth & Pi, 1982) solved this problem by postulating quantum fluctuations in the scalar field, which would "seed" the universe with the slight over-densities (a higher H in Eq. 1 increases the energy density) necessary to precipitate early structure formation. Perhaps the simplest form of fluctuations, scale-free Gaussian fluctuations, appear to satisfy most requirements.

Because the size of the horizon of the Universe during inflation is inversely proportional to the value of the expansion parameter, the large expansion parameter expected during inflation would mean a small horizon. The size of the horizon might remain fairly constant during the bulk of the inflationary stage. With a high inflation rate, newly created quantum-mechanical density perturbations quickly expand to scales in excess of the horizon.

Once the period of rapid inflation ceases, a coasting and decelerating Big Bang-type cosmology is thought to have reasserted itself. With the expansion parameter now declining, the horizon would expand – so fast that it catches up with, and passes perturbations which had once been inflated beyond it. The gravitational clustering of and among relic inflated quantum-mechanical perturbations on scales less than the horizon may then proceed.

When the universe has cooled sufficiently, the Standard Model posits an intricate physical picture of the condensation of fundamental particles, and their coalescence into the neutral matter, termed Standard Big Bang Nucleosynthesis (hereafter SBBN). The part of this picture that concerns us most in this chapter is the relative concentrations of atoms and isotopes formed in the first few minutes following the cessation of the inflationary phase, for these may allow some physical parameters of the model to be constrained by observations of primeval abundances. We briefly describe some of the more conceptual hurdles of this theoretical development because it represents one of the two most compelling connections that the Standard Model has made between cosmological theory and observation.

With expansion, the universe cools. When it is cool enough, matter condenses out of the high-energy photons – neutrons and anti-neutrons, protons and anti-protons in almost equal numbers. Reactions involving protons, neutrons, electrons and electron neutrinos sustain the neutron density in an equilibrium according to the Maxwell–Boltzmann equation. The equilibrium neutron to proton ratio is a function of the temperature of the plasma,

$$\frac{n_n}{n_p} \simeq exp\left(-\frac{(m_n - m_p)c^2}{kT}\right), \tag{2}$$

where the numerator represents the mass difference between the neutron and the proton, expressed in energy equal to 1.29 MeV. As the temperature cools, then, the ratio of neutrons to protons declines. However, due to the expansion of the Universe, the rates of the reactions that sustain the equilibrium drop precipitously, and the neutron–proton ratio becomes "frozen in" at a value of about 0.2. The decay of free neutrons ($\tau_{1/2} \simeq 620$ s) imposes a steady, moderate decline of this ratio.

The next event of importance to us is the formation of deuterium, D, by the fusion of one proton and one neutron, at an age of about 200 s. The amount of D initially formed is proportional to the baryon density. But deuterium is very reactive, and a host of other atoms and isotopes can be formed from it, predominantly ^{4}He. The rapidity of these reactions means that the more D atoms that were initially formed (e.e., the more baryons there were initially), the fewer survive that phase of nucleosynthesis, and the more ^{4}He that is formed. Thus, the deuterium abundance should decline with increases in Ω_b. We are interested in there being a concordance between the ^{4}He abundance and the D abundance, as this would tell us Ω_b (see §5.3.3).

The question of the baryon density raises the question of the baryon–antibaryon asymmetry, for it is by this asymmetry that any baryons exist at all. The current baryon-to-photon ratio represents evidence that to a high accuracy, the number of baryons exceeded the number of antibaryons by a small amount. This asymmetry is nicely encapsulated in the photon to baryon ratio; $\eta = n_\gamma/n_\mathrm{b}$. It is thought that this ratio faithfully expresses the original asymmetry, for the annihilation of antiparticles would produce photons. The only remaining baryons would be due to an excess of one type (baryons) over the other (antibaryons). Furthermore, since this would have occurred during the matter nucleation phase following inflation, any variation in the relative magnitude of the asymmetry could be revealed by observations of isotope ratios. It seems clear, however, that there is little evidence for domains dominated by antibaryons. An identical asymmetry must be invoked to explain the existence of electrons relative to positrons.

Let us return to our now post-nucleosynthesis expanding universe. Within the post-inflationary horizon very small perturbations may initially be damped owing to large particle velocities, and the coupling between baryons and photons. The dark matter, however, (if cool) is less affected by this, and perturbations grow, which tends to expel matter from under-dense regions (with subcritical density) and draw baryons toward over-dense regions (locally supercritical), producing pressure gradients that heat gas to temperatures larger than the norm. These are the acoustic waves that were predicted in the power spectrum of the microwave background.

When the temperature of the Universe cools to about 3500 K, hydrogen becomes increasingly neutral, and the photons begin to "freestream", becoming a relatively uniform background. The wavelengths of these background photons gradually increase at a pace equal to that of the increase in the physical scale of the Universe. Seen from the present, the angular distribution of CMB photons retains the signal of the perturbations in temperature present at the epoch of decoupling. Because of the mode of the expansion of the Universe is determined by basic cosmological parameters, the microwave background data may be used to constrain those parameters, as we shall presently describe.

5.1 The Advent of the Cosmological Constant

The cosmological constant, Einstein's "greatest blunder", has vaulted from a position of astronomical obscurity to one of prominence during the last decade or so. One reason for this is a perceived "awkwardness" between the ages of the oldest Galactic globular clusters, and the age of the Universe as a function of the Hubble constant coupled with the favored cosmologies. An Einstein–de Sitter Universe, a simple model having a critical density, zero curvature and zero pressure, with $h = 0.7$ has an age of 9.3 Gyr; an open universe with $h = 0.7$ and $\Omega_{\mathrm{m}} = 0.3$ has an age of 12 Gyr. The "awkwardness" comes in because the oldest MW globular clusters were thought to range from 14 to as much as 20 Gyr old. In fact, modern isochrones still lead to some globular cluster ages approaching 14 Gyr.

However, it is obvious that globular clusters cannot be older than the Universe they are a part of, and therefore either the assumed cosmological models must be incorrect, or globular cluster ages are really significantly younger. In response to the persistence of this problem, and perhaps also from the rather happy experience with the inflationary paradigm, it was thought that a cosmological constant could help to answer this problem.

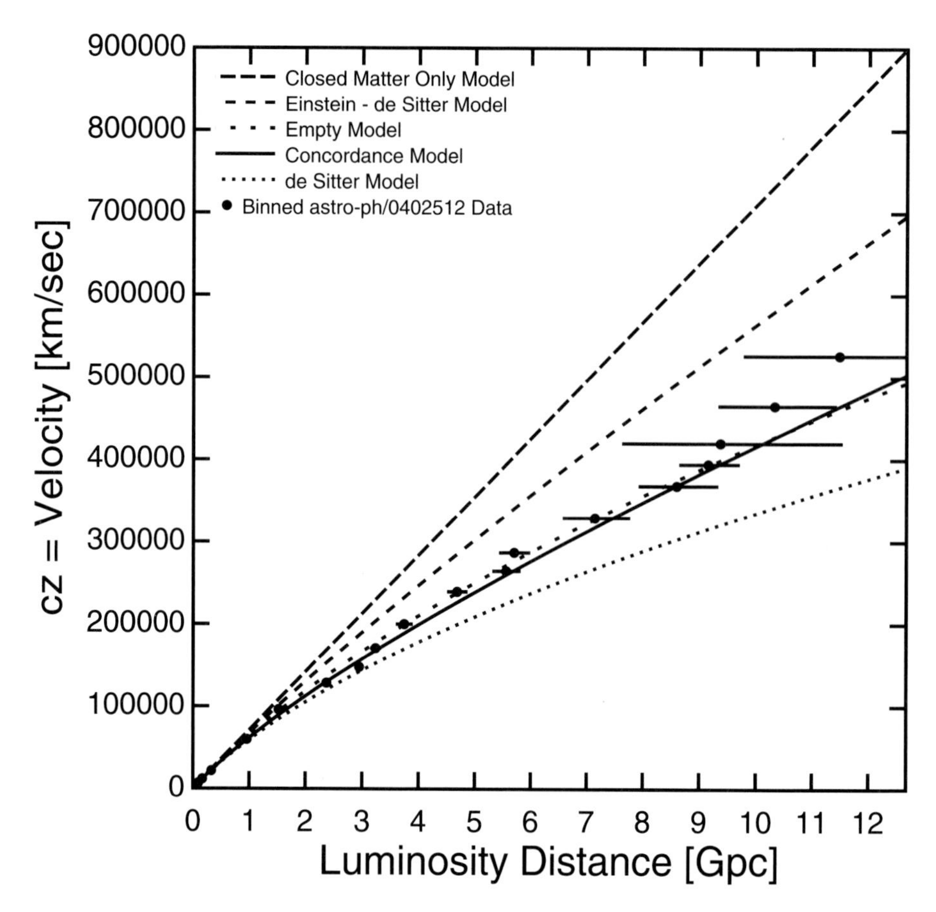

FIGURE 1: Derived luminosity distances of SNe Ia (see Eq. (3). Curves (see labels) show a closed Universe ($\Omega = 2$), the Einstein–de Sitter Universe ($\Omega = 1$) , the empty Universe ($\Omega = 0$), the WMAP-based concordance model with $\Omega_M = 0.27$ and $\Omega_V = 0.73$, and the de Sitter (steady state) model. The concordance model finds that $H_0 = 71$ km s^{-1} Mpc^{-1}, which has been used to scale the luminosity distances in the plot. The figure is courtesy of Ned Wright (http : //www.astro.ucla.edu/ wright/sne_cosmology.html), from a compendium of data from Riess et al. (2004).

The Hubble diagram shows the apparent magnitudes (or the equivalent) of standard candles as a function of redshift. It was initially used with estimates of distances to E-galaxies to derive a possible deceleration parameter with astronomically popular models and data from the era 1960–1980 (with $H_0 = 50$, typically). At that time the ages of stars and the expansion were both thought to exceed 15 Gyr (Sandage, 1961).

More recently, the Hubble diagram has been very effectively used to plot data from high-redshift supernovae type 1a (SNe 1a), which arise from the thermonuclear explosion of carbon–oxygen white dwarfs having a mass equal to the Chandrasekhar mass (Hoyle & Fowler, 1960), 1.4 $M_\odot$. Since SNe 1a at maximum light are extremely bright ($M_B \simeq -19.5$ mag), they were found useful for the measurement of the deceleration parameter, q_0. However, SNe 1a were found to be un-robust "standard candles". An extensive study of low-redshift SNe Ia found that their peak luminosity varied by a factor of $\sim$3. Fortunately, it was found that the decrease in B-band brightness over 15 days from the maximum was a good predictor of its absolute magnitude. Eventually template fitting with multicolor spectra was used to demonstrate a mild scatter, approaching $\sigma \approx 0.15$ mag (*e.g.* Riess et al., 1998).

An example of a SN 1a Hubble diagram is shown in Fig. 1, where data are shown in relation to various cosmological models (see the caption). The horizontal axis is given in units of the luminosity distance. The luminosity distance is defined observationally by the equation,

$$d_{\mathrm{L}} = \left(\frac{\ell_{\mathrm{int}}}{4\pi\mathcal{F}}\right)^{1/2}, \tag{3}$$

where $\mathcal{F}$ is the observed bolometric flux, and ℓ is the intrinsic luminosity of SNe 1a. Given the observed maximum brightness, and the (multicolor) light curve shape, an intrinsic luminosity ℓ can be determined. Then the luminosity distance is calculated as above (Eq. (3)). The luminosity distance is the distance an object would have to be in order to be perceived to have brightness $\mathcal{F}$ in a static flat universe.

Thus, the derived d_{L} values together with the redshift contain the cosmological effects that may allow us to determine in which kind of Universe we live. As one can see from Fig. 1, even with the HST the precision of the determination of the flux at maximum light, and the fitted absolute luminosity, as parameterized by d_{L}, the uncertainty for an individual SN is relatively large; cosmological parameters will be extracted only with the help with a large data-set, which can minimize random (non-systematic) errors.

The results of recent work using a compilation of data from SNe 1a (Tonry et al., 2003) is shown in Fig. 1 suggest that a Universe with a cosmological constant (see solid concordance line in Fig. 1) provides the best fit, though the "empty" model looks about as good. This can also be seen in Fig. 2, where the redshift is now the x-axis, and the y-axis is the variation of the distance modulus relative to an empty universe with no cosmological constant. The individual data points (greyed points and error bars) have been averaged in redshift bins (solid points and error bars). The lines represent a flat Λ model, an open model, and an Einstein–de Sitter model (upper to lower). The distribution of points appears most consistent with the flat Λ model ($\Omega_{\mathrm{m}} = 0.3$, $\Omega_\Lambda = 0.7$).

5.2 "Dark Energy"

The cosmological constant is a repulsive acceleration that was originally proposed by Einstein as a method of stabilizing a static massive universe against gravita-

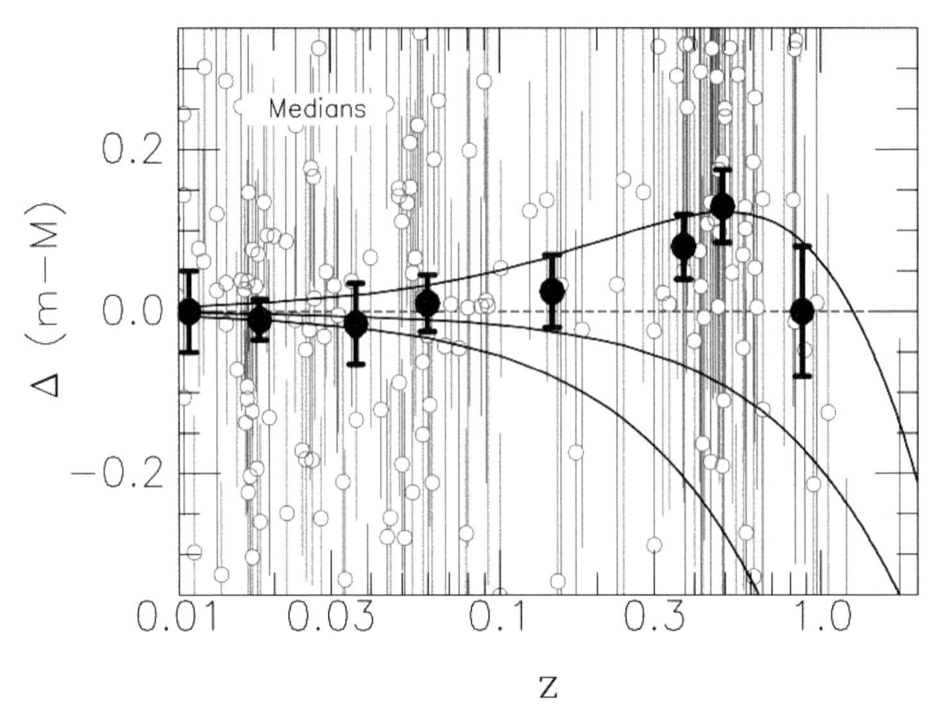

FIGURE 2: Distance modulus residuals relative to an empty expanding universe. When the 170 SNe data are averaged in redshift bins (note solid symbols with error-bars), there appears to be a rather smooth trend. Lines indicate cosmological models defined by $(\Omega_{\rm m}, \Omega_\Lambda)$ =(0.3,0.7), (0.3,0.0), (1.0,0.0). The data show an accelerating Universe at low to moderate redshifts but a decelerating Universe at higher redshifts, consistent with a model having both a cosmological constant and a significant amount of dark matter, which decelerates the expansion through its gravitational action at early epochs ($z \gtrsim 0.5$). Figure from of Tonry et al. (2003).

tional collapse. It fulfilled its purpose, at least theoretically, of stabilizing the static Universe against gravitational collapse. As we have seen, a modern version of the cosmological constant – *inflation* – has been proposed for the very early Universe. Inflation involves a constant expansion parameter, and an insertion of some form of positive energy density (latent heat) to balance the energy it takes to expand the Universe. This has been dubbed a "dark" energy since its physical properties are unknown, other than its tendency to lower the curvature of the Universe. The Friedmann equation for a pure cosmological constant model is

$$H_\Lambda = \sqrt{\frac{8\pi G}{3}\rho_{\rm crit}}. \tag{4}$$

By this equation, if H_Λ is constant, so must be the critical density. The cosmic expansion law $v = H_\Lambda R$ yields an exponential expansion,

$$R = R_0 \exp(H_\Lambda\,(t - t_0)), \tag{5}$$

which implies an acceleration of free-floating laboratories relative to each other. As a consequence, a volume inflates as $V = V_0 \exp(3H_\Lambda(t - t_0))$. Thus, Eq. (4) shows that some energy form must be being created within a given comoving volume while the space is inflating, otherwise ρ_{crit} would decline. If ρ_{crit} declined in time, it would not generate a cosmological constant model. Alternatively, the value of the cosmological constant implies a constant "dark" energy density in the universe. Therefore the expansion of the Universe requires its dark energy content to increase with time ($E(t) = E_0 \exp(3H_\Lambda(\Delta t))$). As dark energy floods space, the fraction of the energy density which is matter would get progressively smaller. What is the future in such a universe? Nagamine & Loeb (2004) performed a speculative numerical simulation showing that structure formation (*i.e.*, the rate of growth of perturbations) would freeze at between 1 and 2 Hubble times in the future, as accretion would be cut off. The gas in the universe would evolve into two distinct phases – a low-density, low-temperature ionized gas cooled adiabatically by the exponential expansion, and a high-density, high-temperature gas in virialized dark matter halos which cools much more slowly by atomic processes. In this model, the volume-filling fraction of over-dense regions must decline exponentially, leaving a smattering of widely scattered "island" clusters. In following this scenario to the far distant future, it is difficult to avoid the speculation that these groups would gradually merge to MBHs that would then evaporate by Hawking radiation (thermal radiation emitted from just outside the Schwartzschild radius of black holes) over immeasurable stretches of time.

In a speculative field such as cosmology, it may be wise to take care with our energetic accounting. If energy is conserved in the Universe, the addition of dark energy to the universe must be accompanied by the removal from the Universe of an equal amount of energy. The only viable method to extract energy appears to be by the addition of a gravitational potential energy, which is negative. The total energy of such a spatially flat universe must be quite close to zero. The negative energy aspect of the cosmological constant must be attributed to the increased cosmological displacements brought about by the accelerating expansion. However, the physical cause of the spatial inflation appears to be as elusive as the nature of the dark energy which is added.

The current most popular explanation for the positive energy density represented by the cosmological constant is *vacuum energy*. Vacuum energy refers to the quantum-mechanical view of the vacuum as a source of virtual particle–antiparticle pairs (see Turner, 2000 for an excellent review of the problem). However, from the standpoint of inflationary theory, the "smallness" of the cosmological constant is a mystery. "Sensible" calculations result in estimates larger by from 10^{55} to 10^{122} times the plausible current value (Weinberg, 1989). That such a vacuum energy density should cause a spatial inflation is not explained in physical terms, but we expect that the dark energy is correlated to an expansion of the universe. In fact, all we know is that there is an accelerating expansion, and that the Universe is spatially nearly flat (the result of recent Wilkinson Microwave Anisotropy Project findings; see §5.3).

The essential difference between the manifestations of an early inflation, and that of a late-time cosmological constant is that with the former the released latent energy is of a density large enough for spontaneous nucleations of particle pairs to occur. For the slow spatial inflation of the present epoch, the energy density of space implied by the Λ-term (currently $\lesssim 10^{-8}\,\mathrm{erg\,cm^{-3}}$) is far too low to allow for direct nucleations without postulating "new physics". Though particle–antiparticle pairs are thought to pop into and out of existence, they are only "virtual" particles; they cannot endure beyond the Heisenberg condition, $\Delta t \leq h/\Delta E$, where h is Planck's constant. Thus, it is thought that the positive energy concomitant with the spatial inflation is locked into a dark energy. This dark energy would act to globally flatten the geometry of the Universe from the hyperbolic geometry it would have with a subcritical mass density. However, this effective mass density is assumed to be incapable of gravitationally clustering.

The more general Friedmann equation for a matter-dominated Universe (compare to Eq. (1)) is,

$$H^2 \equiv \left(\frac{\dot{a}}{a}\right)^2 = \frac{8\pi G}{3}\rho_\mathrm{m} - \frac{k}{a^2} + \frac{\Lambda}{3}, \tag{6}$$

where a is the scale factor of the Universe, ρ_m is the density of matter, and $k = -1$, 0, or 1, according to the three possible geometries of the Universe: open, flat, or closed, respectively. Notice that k must have units of velocity squared. By way of definition,

$$\Omega_\mathrm{m} = \frac{8\pi G \rho_\mathrm{m,0}}{3H_0^2}, \tag{7}$$

$$\Omega_\Lambda = \frac{\Lambda}{3H_0^2}, \tag{8}$$

$$\Omega_\mathrm{k} = \frac{-k}{(a_0 H_0)^2}, \tag{9}$$

where $\rho_\mathrm{m,0}$ is the present-day matter density.

These equations represent the three terms that can contribute significantly to the energy density of the Universe today under the Standard Model of cosmology, and leads to the following redshift-dependent equation for the Hubble function,

$$H(z) = H_0\sqrt{(1+z)^3\Omega_\mathrm{m} + (1+z)^2\Omega_\mathrm{k} + \Omega_\Lambda}, \tag{10}$$

where $(1+z) \equiv (a_0/a)$. For a post-inflationary Universe, it is thought to be a good approximation to assume $k = 0$. Astronomers are not currently interested in non-inflationary models owing to the cosmological problems alluded to earlier. However, a fundamental constraint of the Universe is that its comoving volume should remain constant. If the Universe is a zero-energy (zero curvature) universe, then $\Omega_\Lambda = 1 - \Omega_\mathrm{m}$, where Ω_m must decline as $(a_0/a)^3$. Λ is thus constrained to be less than $3H_0^2$.

This new view of Λ is often expressed in terms of the *equation of state*, in which the pressure is a linear function of the summed energy density,

$$P = w\epsilon, \tag{11}$$

where w is a dimensionless number. Some, however, consider it plausible that w is a function of redshift ($P = w(z)\rho$, where $w(z)$ may be expressed as some linear function). This variant of a Λ model is called *quintessence*. It violates the form of the Friedmann equation (Eqs. (6) and (10)) unless we combine the energy densities comprised by matter and dark energy, but allow for a time-dependent evolution of the energy density that is a mixture of competing rates. In this case, the Friedmann equation becomes,

$$H \equiv \frac{\dot{a}}{a} = \sqrt{\frac{8\pi G}{3c^2}\epsilon}, \tag{12}$$

where $\epsilon = \epsilon_{\rm m} + \epsilon_\Lambda$, $\epsilon_{\rm m} = \epsilon_{\rm m,o}(a_{\rm o}/a)^3$, and $\epsilon_\Lambda = (c^2\Lambda/(8\pi G)$. Redshift is related to scale length by the equation, $a_{\rm o}/a = (1 + z)$. The parameter w now determines the balance between the matter and the vacuum energy density. Under this scheme, $w = -1/3$ is characteristic of a radiation-dominated or relativistic matter-dominated Universe. For $w < -1/3$ there is a component of dark energy. When $w = -1$, there is a cosmological constant. There are no physically realistic equations of state with $w < -1$. For the purposes of this book, we regard quintessence as speculative, and return to the firmer ground of the standard Λ-dominated Friedmann model.

A way to keep a flat Λ model from describing a purely exponential expansion is the fact that the energy density of matter would be declining with time, while the cosmological constant would be, well, constant. The cosmological constant, in the Standard Model, only recently arrived as the dominant actor – i.e., at $z \simeq 0.33$ in the evolution of the large-scale Universe. This is close to the turnover redshift of distance modulus residuals for the flat Λ model in Fig. 2 (top curve). As the Big Bang coasting component of H declines, Λ will become more dominant and Ω_Λ will approach a value of 1. Thus, it is thought that in the far future, our Universe will have a matter density exponentially close to zero and a dark energy equally close to the critical. In the distant future, our local group will move away from Virgo in physical distance, and the accretion remnants of the Milky Way and the Andromeda galaxy will be alone within the event horizon (Nagamine & Loeb, 2004).

5.3 CMB-derived parameters

The observation and analysis of the CMB currently occupies the center stage of observational cosmology.

First recognized in 1964 by Dicke, Peebles, Roll, and Wilkinson after measurements by Penzias & Wilson (1966), the cosmic microwave background (CMB) provides us with strong evidence for a Big Bang cosmology. The observed background had a classic black-body spectrum, with a temperature of 2.73 K (Smoot et al., 1991). The interest in connecting the theory of the Universe with the study of the spatial temperature fluctuations of the CMB has led to several strong astrophysical breakthroughs in the last 10 or so years. The Differential Microwave

Radiometers on the Cosmic Background Explorer (COBE) (e.g., Smoot et al., 1991), provided the first whole-sky look at the minor ($\delta T/T \simeq 3 \times 10^{-5}$) fluctuations with an effective resolution of $\sim 6°$ (see Fig. (3)). It should not go unmentioned that the dipole anisotropy of the CMB showed that our Local Group is moving at a velocity of $\sim$365 km s^{-1}in the direction of $(l, b) = 265°, 48°$, as shown in Fig. 3. The derived peculiar motion of the Galaxy is 547 km s^{-1} relative to the CMB rest frame. This provides a new and important constraint on our understanding of the dynamics and the mass-distribution of our "neighborhood" on scales of $\sim$50 Mpc.

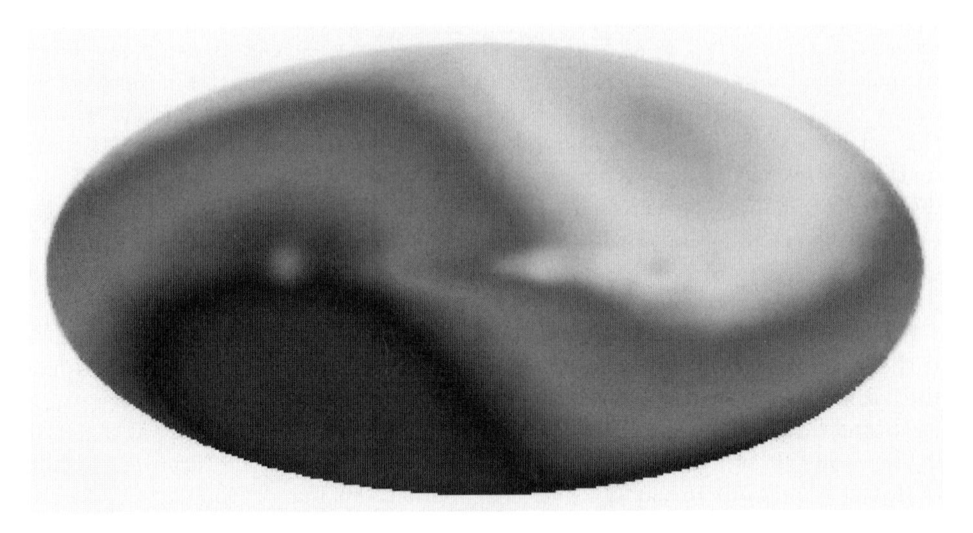

FIGURE 3: The dipole anisotropy, detected with the differential microwave radiometer on COBE (Smoot et al., 1991). The Galactic coordinates $(l, b) = (0, 0)$, is located at the center, and b increased from zero toward the left. The CMB temperature is higher in the upper-right, indicating owing to our motion in that direction (galaxy rotation, local group motion) relative to the cosmic rest frame. (Reproduced in colour in the colour section.)

There followed a series of increasingly precise and generally higher angular resolution instruments which led to the launch of the Wilkinson Microwave Anisotropy Project (WMAP) and the publication of the WMAP first-year conclusions (e.g., Spergel et al., 2003; Kogut et al., 2003; Bennett et al., 2003), regarding the values of cosmological parameters, temperature/polar- ization correlations, and the foreground emission, respectively. Table 1 shows recent determinations of cosmological parameters, where the WMAP values are taken from the WMAP-only data (Spergel et al., 2003) and the CBI (Cosmic Background Imager) polarization data (Readhead et al., 2004).

WMAP results are easily extended by the constraints applied by the addition of independent data from the low-redshift Universe; the correlation lengths of large scale structure (2dF), and supernova studies which probe the $z \lesssim 1$ Universe. Also aiding this enterprise are numerical models of the Universe.

TABLE 1: Cosmological Parameters Constrained by WMAP and CBI

Parameter	WMAP[a]	CBI[b]+WMAP
Ω_{b}	0.047 ± 0.006	$0.04232 \pm .00236$
Ω_{m}	0.29 ± 0.07	0.25 ± 0.06
H_0 $\mathrm{km\,s^{-1}\,Mpc^{-1}}$	72.0 ± 5	74.2 ± 6
Age of Universe τ_0	13.4 ± 0.3 Gyr	13.5 ± 0.3
Re-ionization; z_{re}	17 ± 5	16.0 ± 0.6
Decoupling; z_{dec}	1088^{+1}_{-2}	–
Thickness of last scatter, Δz_{dec}	194 ± 2	–
Sound horizon at decoupling, r_{s}	144 ± 4 Mpc	–
d_{A} to last scattering	13.7 ± 0.5 Mpc[c]	–
Acoustic angular scale[d], ℓ_A	299 ± 2	–
Baryon/photon ratio η	$(6.5^{+0.4}_{-0.3} \times 10^{-10})$	–
Current baryon density, n_{b}	$(2.7 \pm 0.1) \times 10^{-7}$ cm^{-3}	–

[a] Fit to WMAP data only, (Spergel et al., 2003), Table 2.
[b] Cosmic Background Imager (Readhead et al., 2004) constraints on WMAP data.
[c] Comoving distance; proper $d_{\mathrm{A}} = 12.6 \pm 0.4$ Mpc.
[d] $\ell_A \equiv \pi\, d_{\mathrm{A}}/r_{\mathrm{s}}$. This definition is somewhat arbitrary.

The power spectrum $P(\ell)$ is a statistical summary of the angular distribution of the point-to-point spectral fluctuations of the CMB with the "foregrounds" removed. The angles "retained" in the "solutions" range from as much as 40° to as small as a few arcminutes. Instruments such as the Millimeter Anisotropy eXperiment IMaging Array (MAXIMA), Balloon Observations Of Millimetric Extragalactic Radiation ANd Geophysics (BOOMERANG), and WMAP data, which may resolve the surface of last scattering to displacements approaching $10'$. Arrays are being built on the South Pole (e.g., ACBAR) to observe fluctuations on still smaller scales.

The WMAP satellite was hailed as the instrument which would bring precision cosmology. Finally, those protracted battles over the value of the Hubble constant, whether the Universe would collapse or expand forever, or the density of the Universe, would be precisely answered within the context of a self-consistent model of the Universe. Indeed, it has been very successful.

Let's look at some important areas where WMAP (in its first year) appears to have clarified observational cosmology, and a couple of areas where the clarification has not been so well established.

5.3.1 The Curvature of Space: What Acoustic Oscillations Tell us

We have expectations for the spatial scale of regions of high/low mass density immediately following decoupling – based on the physical scales of large scale structure at the present time and the redshift at decoupling. The Cosmological Standard Model will then determine what angular displacement at the decoupling epoch corresponds to these spatial scales. But what signature would such density perturbations have left? Because the universe before this time was ionized, the early radiation field is coupled to the baryonic field. A result of this coupling is that over-dense regions have "harmonic" frequencies, leading to acoustic oscillations, which show up as maxima in the power spectrum of the CMB and its small fluctuations. The largest perturbations within the horizon at this epoch may form compressive hot-spots in the CMB due to the convergent motions of baryons responding to gravity and radiation pressure gradients. Smaller perturbations, which formed earlier*, may display a rarification, following a still earlier initial compressive stage. Yet smaller perturbations may be observed experiencing their second compression, and so on. Since perturbations of a given mass-scale correspond to a redshift at which they entered the horizon, there will be a characteristic angular scale at which the point-to-point correlations will experience increased power.

The strength of this signal is largely due to the fact that kinetic action (e.g.,, sound speed) proceeds at a significant fraction of the speed of light ($c_s = c/\sqrt{3}$). The position of the first (latest) peak in the power spectrum of CMB temperature perturbations should represent the angular scale of structures undergoing their first compressive acoustic maximum within the horizon during decoupling. This value can constrain the current matter density of the Universe.

The method by which this is done is complicated by the methodology of the analysis of the temperature fluctuations. The fluctuations are fit to spherical-harmonics. This angular power-spectrum is parameterized in terms of ℓ, the spherical harmonic multipole number, which refers to an angle $\theta \sim 200^\circ/\ell$. A fully reduced image of the CMB temperature fluctuations is shown in Fig. 4 (Bennett et al., 2003). Though over-processed for scientific use, its appearance probably accurately conveys the impression of small-scale fluctuations permeating the fabric of much larger-scale perturbations viewed from beyond the Galaxy. Notice the rather large-scle fluctuations at angles of from 20° to 40°. There appear to be some large-scale circular maxima or minima, especially near the Galactic equator. The reader should recognize, however, that this projection of the whole sky introduces significant distortions of the shape of regions when they lie closer to the edge of the map.

When the power of the spherical harmonics are plotted as a function of ℓ, there appear a sequence of peaks as ℓ increases (i.e., as the angle *decreases*). The value of ℓ at which the first peak is a maximum is indicative of the cumulative effects of the rate of expansion of the Universe following decoupling, and hence of the combined effects of the matter density and the (possible) cosmological constant. These effects are parameterized into a measure called the *angular diameter distance*, d_A, which is the distance at which the proper displacement d at high redshift is determined by the equation $d = d_A \sin\theta$, where θ is the measured angle across the object. The spherical multipole wavenumber that corresponds with this is $\ell \simeq 200/\theta$, when θ is given in degrees. Thus the first acoustic peak should occur at $\ell \propto d_A$.

---

*Recall that following the end of inflation, the Universe entered a long phase of progressively larger perturbations entering a progressively larger horizon.

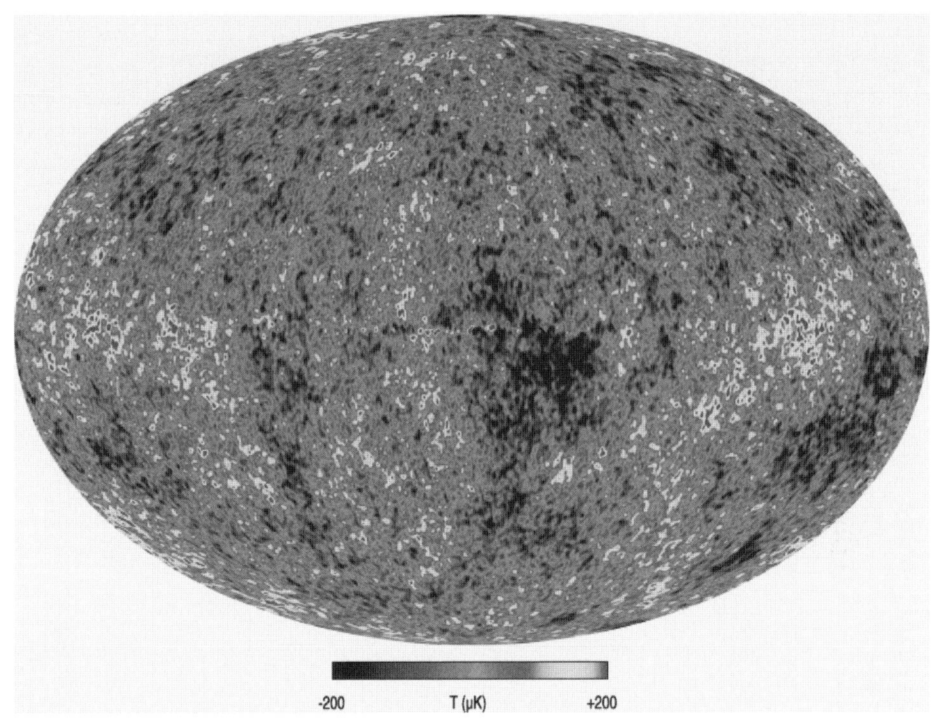

FIGURE 4: . "Internal linear combination" map combining the five band maps of the WMAP data in such a way as to subtract "foreground" contaminating fields, such as various features of our Milky Way galaxy, while constraining the mean CMB flux to be constant. For a more detailed description, see Bennett et al. (2003). The plane of the galaxy is horizontal, the center is at a longitude of 0 degrees. (Reproduced in colour in the colour section.)

The analysis of the CMB in WMAP shows that the first peak appears at $\ell \simeq 220$, referring to an angle slightly less than $1°$. If the Universe is over-dense ($\Omega \gg 1$), $d_{\rm A}$ is small, ℓ is small and $\theta_{\rm first} \propto d_{\rm A}^{-1}$. For instance, for a redshift of decoupling $z_{\rm dec} = 1088$, a matter density $\Omega = 2$ ($\Lambda = 0$), and $h = 0.7$, then $d_{\rm A} = 3.3$ Mpc, while for $\Omega = 0.3$, $d_{\rm A} = 20.2$ Mpc. Seen in this way, it is clear that the position of the first acoustic peak is sensitive to the matter density, and indicative of the overall geometry of the Universe. It may be used to constrain $\Omega_{\rm m}$.

Since the first peak represents the maximum compression of perturbations that are collapsing for the first time, the second peak should represent perturbations at $\ell \simeq 500$, or angular scales of $\theta = 0.4°$ which entered the horizon at an earlier time, and that are seen, at the moment of recombination, experiencing their second compression. The position and power in this perturbation is sensitive to the baryon density since oscillations after the first compression are driven by pressure gradients (which dark matter cannot "feel") resulting from photo-ionization heating. The baryon density $\Omega_{\rm b}$ parameterizes the inertial drag in the response of a cloud of baryons to pressure gradients. Thus, the position of the second peak would be at larger ℓ if Ω_b is larger. This can be seen in Fig. 5, which shows several simulations with different values of Ω_b and Ω_m (see legend). Note that for large Ω_b (the red curves have $\Omega_b = 0.05$ and 0.06, while the blue curves have a low baryon density, $\Omega_b = 0.02$ and 0.03. Note the displacement of the second peak to larger ℓ (smaller θ for the red curves), as one would expect from the inertial restraint on expansion in high Ω_b models.

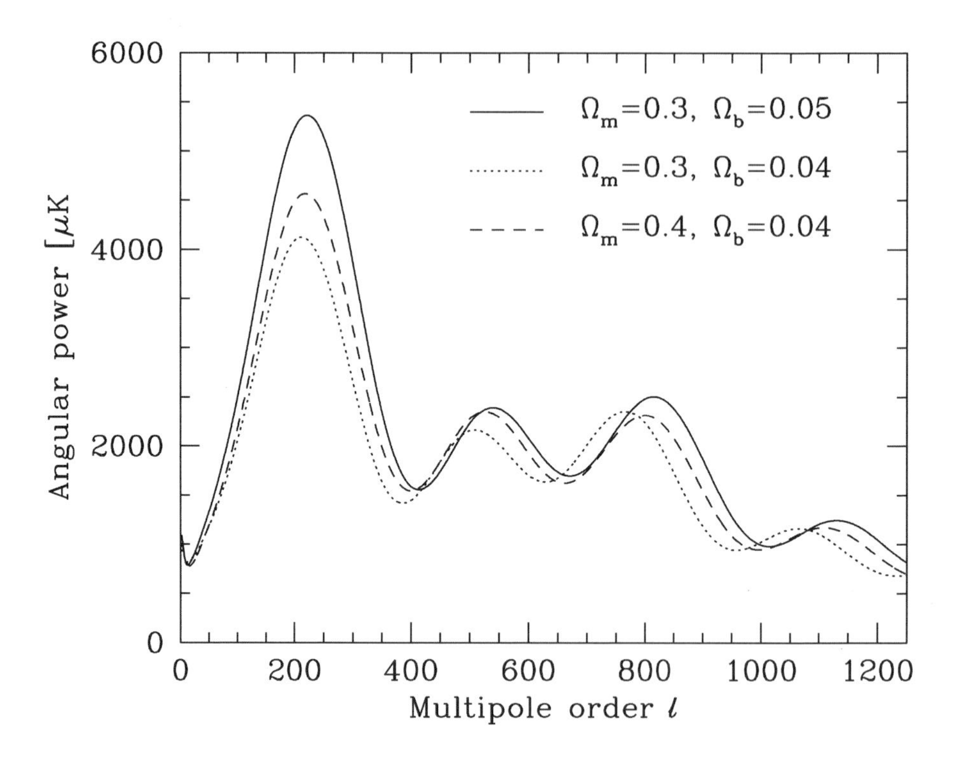

FIGURE 5: The effect of varying baryon and DM density on the CMB power spectrum. Note the displacement toward lower ℓ when matter density is higher (i.e., $d_A(z_r)$ is smaller), and a displacement of the second acoustic peak toward the larger ℓ when baryon densities are larger. The Omega baryons is the number following the "b", while the dark matter density is the number following the "CDM". The simulations all use $H_0 = 70 \quad \mathrm{km\,s^{-1}\,Mpc^{-1}}$. Image from http : //www.physics.ucsb.edu/ jatila/astro/astro2/CMBFAST.html. (Reproduced in colour in the colour section.)

5.3.2 The Correspondence between Large-Scale Structure and $P(\ell)$

Once the parameters that describe cosmology are preliminarily constrained by the analysis alluded to above, we may construct the correspondence between the angular scale of perturbations in the CMB and parameters extracted from the physical structures into which they have grown in the low-redshift universe. We parameterize this structure in terms of correlation lengths of galaxies and galaxy clusters.

Both the supernova projccts, the High-Z Supernova project (Riess et al., 1998), and the Supernova Cosmology Project (Perlmutter et al., 1999; Lidman, 2004), suggest the existence of a substantial cosmological constant; a repulsive acceleration for objects separated by cosmological distances. Both projects concluded that the distant SNe were fainter than anticipated, and luminosity distances were more consistent with a universe having a significant cosmological constant component.

Results from WMAP (Spergel et al., 2003) suggest that the Universe is effectively flat ($k \simeq 0$), which is to say that the mass-equivalent of the energy density of the Universe Ω_0 is close to unity. Together with the rather small matter density ($\Omega_{\rm m} \simeq 0.27$) this implies that that Ω_Λ, the so-called " dark energy" associated with the cosmological constant, clearly dominates the total positive energy density in the Universe.

Combining the WMAP data with other data sets provides an opportunity to break many parameter degeneracies* in the WMAP data. Most prominently, supernova data, which can constrain the luminosity distance $d_{\rm L}$ (see Eq. (3)), is complementary (i.e., degenerate along a different locus) to WMAP data, which is sensitive to the angular diameter distance $d_{\rm A}$. The perturbation spectrum derived from the 2dF galaxy redshift survey (Lahav et al., 2002) may be combined with WMAP data to constrain WMAP parameters (Spergel et al., 2003). Observations of the X-ray gas mass fraction in rich galaxy clusters over a range of redshifts allows additional independent constraints on the dark energy density to be placed if one assumes the gas mass fraction is constant (Allen et al., 2004). The method by which independent data are applied to derive the best-constrained cosmological parameters is nicely presented in Bahcall et al. (1999). This work suggests that the Universe is flat to within the limits of detection, and $\Omega_\Lambda \simeq 0.73$.

Given the optimal fit to cosmology, numerical simulations show that density fluctuations of comoving wavelength ~ 2 Mpc (galaxy "seeds") would subtend an angle $\theta \simeq 30$ arc-sec on the surface of last scattering, while protoclusters which today have clustering scale-lengths of about ~ 20 Mpc, should be seen in the fluctuations at angular separations of $\theta \simeq 5$ arcmin, just barely detectable by MAXIMA (Spergel et al., 2003). The largest structures observed in the local Universe, such as the "Great Wall" ($\sim 100^+$ Mpc; Geller & Huchra 1990) would subtend an angle on the surface of last scattering approaching $\sim 1°$. Our high-z optical observations suggest that much of the luminous matter was distributed in small, sub-galactic systems (see Chapter 3). As we sample lower redshifts, progressively

*This term loosely refers to loci in "parameter space" that are very poorly constrained, such as the age–metallicity degeneracy in stellar astronomy.

larger galaxies are seen, the products of regular accretion and cannibalism.

5.3.3 Constraints on Hubble's Constant and the Age of the Universe

With our new-found confidence in the value of cosmological parameters, we can assign ages or look-back times to objects we see at any given redshift. Our understanding of the evolution of galaxies and the Universe can now be tested for consistency with a more precise time reckoning.

The age as a function of redshift for a flat matter-dominated Universe with a Λ term is (Scott et al., 2000)

$$t(z) = \frac{1}{H_0} \int_z^{\infty} \frac{dz'}{(1+z')\sqrt{\Omega_0(1+z')^3 + \Omega_\Lambda}}, \tag{13}$$

but the integration from $z = 0$ to ∞ gives the age of the universe today,

$$t = \frac{2}{3H_0} \frac{1}{\sqrt{\Omega_\Lambda}} \ln \left\{ \frac{1+\sqrt{\Omega_\Lambda}}{\sqrt{\Omega_0}} \right\}. \tag{14}$$

Perhaps the most important single parameter derived from the recent advances in observational cosmology is the Hubble constant. The cosmological fit provided by the data referred to above (Table 1) finds the Hubble constant to be $h \simeq 0.71$, $\Omega_{\rm m} \simeq 0.27$, and $\Omega_\Lambda \simeq 0.73$. The HST Key Project on the value of H_0 was primarily based on the Cepheid period–luminosity relation applied over distances from 60 to 400 Mpc (Freedman et al., 2001). For the Cepheid calibration alone they found $H_0 = 71 \pm 6.3\ {\rm km\,s^{-1}\,Mpc^{-1}}$, in agreement with the WMAP value. Morover, when combined with a number of other distance measures (SNe Ia and II, Tully–Fisher relation, surface brightness fluctuations, and fundamental plane for E galaxies), their result is $H_0 = 72 \pm 8\ {\rm km\,s^{-1}\,Mpc^{-1}}$. Obviously, the accurate and precise assessment of H_0 is a task of huge proportions. The value extracted from WMAP is based on a concordance model which combines a wide range of observational data together with the Friedmann cosmological model. This allows a precise determination of H_0 and other cosmological parameters whose locally derived values were uncertain. Much, however, depends on the correctness of the basic cosmological model. Before we completely embrace this model, let us look in detail at what the derived parameters are telling us, and whether they are consistent with other observations.

5.3.4 Detecting Polarized CMB Photons in Acoustic Waves

The detection of a polarization signal in the CMB with WMAP data (Kogut et al., 2003) was predicted on the basis Thomson scattering of baryons in acoustic waves when opacities are of order unity. Opacities of unity occur during recombination as well as re-ionization. Because of the symmetry of Thomson scattering, only the quadrupole component of the radiation field that scatters from free electrons produces the polarization. Thus, at the surface of last scattering, the main source of quadrupole radiation is the movement of the baryons within the acoustic waves of matter in gravitational potential wells. Because of this, the maximal effect is expected to occur at a multipole $\ell \simeq 300$, which lies between the first power spectrum maximum at $\ell = 200$ and the next minimum at $\ell \sim 400$ (compressive and rarifactive, respectively). The WMAP polarization data of Kogut et al. (2003) has been recently confirmed by observations with the Cosmic Background Imager (CBI) (Readhead et al., 2004), which is a ground-based array with 13 microwave antennas that can resolve structure down to $\ell \sim 1960$. Some of their results are found in Table 1.

5.3.5 Checking on Big Bang Nucleosynthesis

The "concordance" fit of Spergel et al. (2003), which uses 2dF clustering and SN 1a data, yields $\Omega_b h^2 = 0.0224 \pm 0.0009$ and a matter density $\Omega_m h^2 = 0.135 \pm 0.009$. Given the Hubble constant, $h = 0.71 \pm 0.04$, the cosmology is specified. Given these values, SBBN can predict cosmochemical parameters which may be compared with observations. The two constraints we discuss here are the deuterium to hydrogen ratio D/H (that is, the ratio of their respective number densities), and the mass fraction of helium Y_p.

Recall now the opening comments of this chapter in which the subtle physical routes for the synthesis of D and He were discussed. Observing the D/H ratio is possible by virtue of the slightly different resonant frequency of the deuterium Lyman lines from those of H. Thus, the deuterium component may be resolved in the flanks of the H-Lyman lines in high-resolution spectra of high redshift Lyman limit systems*. A recent paper (Kirkman et al., 2003), citing improved techniques, derives the value $D/H = (2.42^{+0.035}_{-0.025}) \times 10^{-5}$ from a $z = 2.526$ absorption system in the spectrum of a quasar, Q1243+3047. An average with five other previously analyzed absorption systems gives $D/H = (2.78^{+0.042}_{-0.038}) \times 10^{-5}$. Under Standard Big Bang Nucleosynthesis (SBBN), these are consistent with a photon to baryon ratio of about $\eta = (5.9 \pm 0.5) \times 10^{-10}$, in good agreement with the WMAP-derived value $\eta = (6.5^{+0.4}_{-0.3}) \times 10^{-10}$.

The ^4He abundance is parameterized by the primordial helium mass fraction, Y_p. Its value is observationally estimated by noting the limit of the trend in the mass fraction of helium in chemically young galaxies as a function of declining metallicity (Izotov & Thuan, 2004). They find that $Y_p = 0.2421 \pm 0.0021$, which consistent (under SBBN) with $\Omega_b h^2 = 0.012^{+0.003}_{-0.002}$, but smaller at the 2-$\sigma$ level than the WMAP and D/H measurements would indicate. The derived baryon to photon ratio based on the ^{4}He/H observations is $\eta = 3.4 \times 10^{-10}$, almost half the WMAP value.

The situation can be seen graphically in Fig. 6, where the vertical gray bar shows the WMAP results. The deuterium measurements agree very well with the WMAP data. For the other atomic species, however, the fit seems worse. The ^{3}He value is an upper limit (it would "slide" to the right for a lower number fraction). The ^{4}He measurements are more than 2-σ out of agreement with WMAP data on the low η-side, and ^{7}Li also appears discordant.

---

*Lyman limit systems are neutral hydrogen absorption systems that have H I column densities $N_{HI} > 1.6 \times 10^{17}$ cm^{-2} (an optical depth $\tau \gtrsim 1$ at $\lambda_0 \leq 912$ Å.)

Whether these results are a serious concern, or whether some systematic error in the ^{4}He number fraction is the cause of the discordant result, remains to be seen.

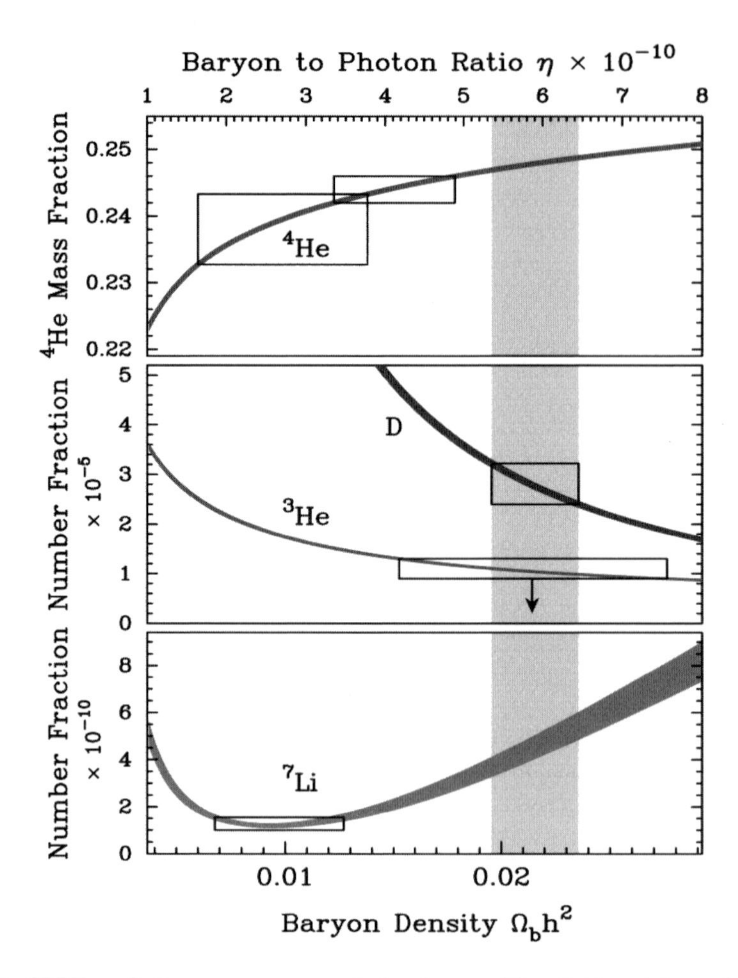

FIGURE 6: SBBN predictions versus measured abundances of four light nuclei as a function of the baryon density. The figure has three vertical panels each with a different linear scale. The top curve is the ^{4}He mass as a fraction of the mass of all baryons, while the three lower curves are the number fractions D/H, ^{3}He/H, and ^{7}Li/H. The vertical widths of the curves show the uncertainties in the predictions. Each of five boxes show measurements, the vertical extension of which is the 1-σ random error. The horizontal position has been adjusted to overlap the prediction SBBN curves. The larger ^{4}He box is from Olive et al. (1997), and the smaller one is from Izotov & Thuan (1998). The new value of Izotov & Thuan (2004) is in between the two He boxes. The D box (middle panel) is the mean from five QSOs from Kirkman et al. (2003). The ^{3}He from Bania et al. (2002) is an upper limit. The ^{7}Li data is from Ryan et al. (2000). We anticipated that all the data boxes should overlap the vertical band that covers the D/H data. That they do not may be because of systematic errors. Figure courtesy of Kirkman et al. (2003).

5.3.6 An Intriguing Discrepancy

Spergel et al. (2003) announced an intriguing discrepancy between the observed angular correlation of the CMBR at large angles ($\theta \gtrsim 45°$; $\ell \lesssim 5$) and the predictions of the best-fit ΛCDM. The WMAP found no significant correlation signal on angular scales exceeding 60° ($\ell \lesssim 3.5$), while the Λ model predicts a significant anti-correlation. It must be acknowledged that the discrepancy may be the result of cosmic variance or foreground contamination; but it may be a genuine problem. As *real*, this discrepancy, at such large angular scales, could be a sign of the new physics entailed by a cosmological constant/dark energy (Spergel et al., 2003) – which could conceivably lead to a constraint on the evolution of the Λ cold dark matter cosmology. Of course, it might not be a statistically significant astrophysical result.

5.3.7 The Ages of the Oldest Galactic Stars

It is of interest to compare the cosmic age of the Universe with the ages of the oldest Galactic (i.e., MW) stars. One always hoped that the stellar ages were advocated independently from the time-cales that emerge from cosmological studies. The cosmic timescale was initially the inverse of the Hubble constant. For the currently embraced value $H_0 = 72\ \mathrm{km\,s^{-1}\,Mpc^{-1}}$, a Hubble time is 13.58 Gyr (as compared to the Universe's age under these parameters, $T_\mathrm{U} = 13.4$ Gyr).

The comparison of the age-dating techniques for the oldest stars and the Universe makes for interesting reading. A decade or two ago this area of inquiry was dominated by projects to determine the value of the Hubble constant. There arose two groups whose respectively discordant values for H_0 provided some amusement, as their favored values differed by a factor of 2!

Sandage & Tammann (1982, 1993) in particular were strong advocates of the "long distance scale" represented by an $H_0 \simeq 50\ \mathrm{km\,s^{-1}\,Mpc^{-1}}$ value. For this Hubble constant and the then-popular open Universe models ($\Omega_\mathrm{tot} \simeq 0.1$) the expansion age came out to near 18 Gyr and few stellar observations and models suggested competitive (large) ages. However, a flat ($\Omega_\mathrm{tot} = 1.0$) model with $H_0 = 50\ \mathrm{km\,s^{-1}\,Mpc^{-1}}$ would be back at an expansion age of near 13 Gyr, and was deemed too competitive with the extant stellar ages given by the globular cluster turnoff luminosities and stellar interior models.

On the other hand, de Vaucouleurs (1982) and observing teams led by Aaronson, and by Tonry (Aaronson et al., 1986; Tonry & Schneider, 1988), favored a shorter distance scale; closer to $H_0 = 100\ \mathrm{km\,s^{-1}\,Mpc^{-1}}$. Thus an expansion age of $T \leq 10$ Gyr was demanded and here again the ages of the galactic globular clusters were clearly too great. It was in response to this demand that the cosmological constant began to be seriously talked about again.

Fortunately, in the intervening years the availability of the HST to image fainter and less-crowded Cepheids with higher fidelity has enabled a gradual convergence of measures of the Hubble constant to a value not far from $70\ \mathrm{km\,s^{-1}\,Mpc^{-1}}$ (e.g., Freedman et al., 2001). Obligingly, the popular ages of Galactic globular clusters have declined somewhat so that the age of the Universe and the age of the oldest clusters are now both about the same.

However, new results from the detailed modeling of the evolution of low-mass stars ($M_* \lesssim 1\,\mathrm{M}_\odot$) (Imbriani et al., 2004) suggest that the age of Galactic globular clusters is ~14 Gyr! This finding comes as the result of an experiment that studied the speed of the carbon–nitrogen–oxygen (CNO) cycle in stars. The speed of the CNO cycle is determined by the slowest of its sub-reactions. In particular, the temperature-sensitive rate of the conversion of H to He by catalysis with the CNO cycle, is limited by the reaction $^{14}N(p,\gamma)^{15}O$. This step has recently been measured at the Laboratory for Underground Nuclear Astrophysics (LUNA) in Gran Sasso, Italy (Formicola et al., 2004). While the CNO cycle is usually thought of as relevant only to massive stars, during the transition from the main sequence to the helium-burning stage, the CNO cycle is activated inside moderately low-mass stars (~0.8 $\mathrm{M}_\odot$). The reaction rates derived by the LUNA experiment, when incorporated into stellar synthesis code, shows that the lower $^{14}(p,\gamma)^{15}O$ rate implies brighter and bluer turnoff points for stars. Thus, stars were being interpreted as being more massive than actuality, and hence younger than actuality. This new work suggests globular clusters are between 0.7 and ~ 1 Gyr older than previously thought.

Independent photometric work on the oldest Galactic globular clusters (Gratton et al., 2003) derived an isochron-modeled age of 13.4 ± 1.0 Gyr *before* the Formicola et al. data was released. Chaboyer (2002) finds the average age of 17 Galactic globular clusters to be 12.6 Gyr old. The average of this and the Gratton value plus the Formicola correction yields an age of 14 Gyr, but with a significant uncertainty. Meanwhile, to reemphasize, the WMAP age of the Universe is 13.7 ± 0.2 Gyr.

A study by Pasquini et al. (2004) shows that beryllium in globular clusters is a good chronometer for age of the globular clusters. Assuming a 200-Myr gap between the formation of the Universe and the formation of the Galaxy, they find that the globular cluster must have been formed between 13.2 and 13.3 Gyr ago. This method, however, implicitly assumes that the first local enhancements of CNO and Be occurred at a Universe age of 200 Myr ($z \sim 17$). Since the cosmic ray production is assumed proportional to the SN rate, and this is assumed to be proportional to the SFR, then the Be content of newly formed stars ought to be proportional to the metallicity Z. The age at which the globular cluster formed is then 200 Myr plus the Galactic SFR at that time times some factor which relates the SFR to the Be production rate. For the legitimacy of this step, we are refer to Valle et al. (2002).

The possibility that stars exist that are older than the expansion age of the Universe would be shocking news to most astronomers. Let us consider in more detail the significance and possible repercussions of this new information. My personal opinion would have been to expect a cosmic age (based on cosmological parameters, h, Ω_m, and Ω_Λ) that exceeds the ages of the remaining few oldest galactic stars by 1 to 2 Gyr (i.e., $z_\mathrm{form} \simeq 3$ to 6). The new result does not leave much breathing room between the two procedures; the formal errors in the respective age determinations (i.e., $T_U = 13.7 \pm 0.2$ Gyr versus $T_{\rm stars} \gtrsim 14.1$ to 14.4 Gyr) are still moderately large.

But even utilizing the large uncertainty in globular cluster ages seems to offer only scant hope of providing the expected 1–2 Gyr buffer between the Universe's formation and the first MW globular clusters. The epoch of rapid galaxy formation appears to start at $z \gtrsim 5$, but, as noted in Chapter 3, most luminous systems at large z are significantly smaller than the present MW. However, if we use $z = 5$ as an arbitrary upper limit on the age of the formation of the MW, then under the WMAP fit of cosmological parameters, the current age of the stars would be 12.5 Gyr old, having formed when the Universe was 1.2 Gyr old. Most scenarios for globular cluster formation utilize either a scaled-up molecular cloud model, or super star clusters formed by mergers. One imaginative proposal (Manning, 1999) uses early sub-galactic perturbations (i.e., Lyα clouds) as proto-clusters. This dissipational scenario involves the accretion of clouds to proto-galaxies, resulting in compression and eventual efficient star-formation. The timing of these events should correspond with the period of rapid decline in the line density of Lyα clouds – somewhere in the redshift range of $z \simeq 2$ to 4. In this case, the age of stars within Galactic globular clusters would be less than 12.1 Gyr; they would have formed when the universe was more than ~ 1.6 Gyr old.

Stellar evolution models are the method by which observations of stars are made into age estimates. Among the sources of uncertainty in main sequence lifetimes, is atomic diffusion (the settling of heavier atoms), the treatment of convection and convective "overshoot", which affects colors, and the mass of hydrogen to be burned on the main sequence. In addition, contemporary model atmospheres must be used to transform the calculated luminosities into observed magnitudes and colors. With this complexity, we may expect further changes in the future.

There are other means of dating the oldest stars, which yield somewhat different values. For instance, the age based on the radioactive decay of uranium and thorium seen in the spectra of very metal-poor halo stars yields a stellar/galactic age of 14 ± 2.4 Gyr. The faintest white dwarfs from the halo cluster M4 yield an age of 12.7 ± 0.7 Gyr (Hansen et al., 2002). The point to be taken home is that the ages of Galactic globular clusters are not known to sufficient accuracy to definitively say that there is an age awkwardness, but the data suggests that we ought to be concerned about such a possibility.

5.4 Summary

The observational cosmology of the last decade has provided the first sense that, with a diversity of approaches, it will be possible to constrain the values of cosmological parameters to a respectable precision. It is unquestioned that the analysis of the CMB has provided the main breakthrough in establishing a preferred cosmology – a flat Λ model with $\Omega_{\rm m} \simeq 0.27$ and $h \simeq 0.71$. These values came about by combining CMB data with low-z galaxy clustering data, and SN 1a luminosity distance data (see §5.3.2). With these values, the equations of SBBN may be run, deriving primordial abundances of various elements and isotopes. Again, see Fig. 6; the observational value of D/H is found to be in good agreement with the concordance cosmology; however, the observational value for $Y_{\rm p}$ is more than 2-σ lower than the value suggested by the concordance model. The other discrepancy that may be of interest is the very low value of the angular correlation function for angular displacements of between 45° and 270° in relation to the predictions of the best-fit ΛCDM model. If real, this disparity may offer clues as to the nature of the expected dark energy.

REFERENCES

Aaronson, M., et al., 1986, ApJ, 302, 536

Allen, S. W., et al., 2004, MNRAS, 353, 457

Bahcall, N. A., et al., 1999, Science, 284, 1481

Bania, T. M., et al., 2002, Nature, 415, 54

Bennett, C. L., et al., 2003, ApJS, 148, 97

Chaboyer, B. 2002, in Proceedings of the 37th Recontres de Moriond, p.37

de Vaucouleurs, G. 1982, Nature, 299, 303

Formicola, A., et al., 2004, Physics Letters B, 591, 61

Freedman, W. L., et al., 2001, ApJ, 553, 47

Geller, M. & Huchra, J. 1990, Scientific American, 262, 19

Gratton, R. G., et al., 2003, AAp, 408, 529

Guth, A. H. 1981, Phys. Rev. D, 23, 347

Guth, A. H. & Pi, S.-Y. 1982, Physical Review Letters, 49, 1110

Hansen, B. M. S., et al., 2002, ApJL, 574, L155

Hoyle, F. & Fowler, W. A. 1960, ApJ, 132, 565

Imbriani, G., et al., 2004, AAp, 420, 625

Izotov, Y. I. & Thuan, T. X. 1998, ApJ, 500, 188

Izotov, Y. I. & Thuan, T. X. 2004, ApJ, 602, 200

Kirkman, D., et al., 2003, ApJS, 149, 1

Kogut, A., et al., 2003, ApJs, 148, 161

Lahav, O., et al., 2002, MNRAS, 333, 961

Lidman, C. 2004, The Messenger (ESO), 118, 24

Manning, C. V. 1999, ApJ, 518, 226

Nagamine, K. & Loeb, A. 2004, New Astronomy, 9, 573

Olive, K. A., et al., 1997, ApJ, 483, 788

Pasquini, L., et al., 2004, AAp, 426, 651

Penzias, A. A. & Wilson, R. W. 1966, ApJ, 146, 666

Perlmutter, S., et al., 1999, ApJ, 517, 565

Readhead, A. C. S., et al., 2004, Science, 306, 836

Riess, A. G., et al., 1998, AJ, 116, 1009

Riess, A. G., et al., 2004, ApJ, 607, 665

Ryan, S. G., et al., 2000, ApJL, 530, L57

Sandage, A. 1961, *The Hubble Atlas of Galaxies* (Washington D.C.: Carnegie Institution of Washington)

Sandage, A. & Tammann, G. A. 1982, ApJ, 256, 339

Sandage, A. & Tammann, 1993, ApJ, 415, 1

Scott, D., et al., 2000, in *Allen's Astrophysical Quantities*, ed., Cox, N., Springer-Verlag, New York

Smoot, G. F., et al., 1991, ApJL, 371, L1

Spergel, D. N., et al., 2003, ApJS, 148, 175

Tonry, J. & Schneider, D. P. 1988, AJ, 96, 807

Tonry, J. L., et al., 2003, ApJ, 594, 1

Turner, M. S. 2000, Physica Scripta Volume T, 85, 210

Valle, G., et al., 2002, ApJ, 566, 252

Weinberg, S. 1989, Reviews of Modern Physics, 61, 1

Chapter 6

Astronomical Instrumentation of the Future

6.1 Of Needs and Strategies

Science advances through well-thought-out concepts, surprises, and new instruments. In this chapter we discuss several promising new instruments on the time horizon; their maturity may lead to dramatic advances in the study of galaxies in their formative stages. For example, what will be the active and important instruments to take us toward the epoch of galaxy formation? (A philosophical question might be: Is there an "epoch" of galaxy formation, or is there merely an epoch of earliest galaxy formation?)

We are hoping to study the IGM through higher spectral purity observations of distant, moderately bright QSOs. An example would be the advanced ESO spectrometer CRIRES, a high-resolution Cryogenic IR Echelle Spectrometer ($\lambda \sim$1–5 μm range), to be installed at the VLT (ESO's Very Large Telescope) in 2005. A planned IR spectrometer for high-resolution observations on the Keck II telescope is "possible" (Keck IR Multi-Object Spectrograph) – a multi-slit cryogenic imaging spectrograph that would fill a similar need. Potential science from these spectrometers might be the study of intervening Mg II absorption systems at $z \gtrsim 3.3$.

The instruments being built or in the planning stage will have enhanced capabilities to attack the selection of very distant young galaxies which, because of their redshift, radiate fairly exclusively in the infrared (or even the sub-millimeter) part of the spectrum. Also some other galaxies are copious IR emitters as their shorter wavelengths are absorbed and re-radiated by warm dust.

A common thread in our needs in the near future is the push toward high angular resolution observations at long wavelengths. That means, for ground-based optical observations, using adaptive optics (AO). For sub-millimeter to radio interferometry, we require sites with unusually dry air – the greater the altitude the better. To advance in these sample areas it will require either space-based instruments or ground-based instruments able to correct for the effects of variable atmospheric turbulence and moisture. For instance, we anticipate excellent IR data from the orbital Spitzer (formerly SIRTF) satellite, with photometric capability from the near- to mid-infrared region, and the James Webb Space Telescope (JWST), a futuristically planned 6-m IR telescope destined for a distant (the L2 Lagrangian point) solar orbit. It should be launched in about 2011 into an orbit with excellent freedom from Earth-bound foreground radiation.

Even larger ground-based telescopes may go into operation on roughly the same timescale. The Carnegie Institution and the University of Arizona and others, are planning a Giant Magellan Telescope, or GMT, whose 7–8.4-m mirrors are the equivalent of a single 20-m telescope. Under current study is a 30-m concept being explored by a joint effort of the University of California, Caltech, the National Optical Astronomy Observatory, and likely colleagues from Canada. A European consortium is exploring a 100-m telescope – the OWL (OverWhelmingly Large telescope). We discuss these in more detail in §6.3.1.

These all-purpose giant eyes to the sky will have many virtues when their locations are favorable and their IR technology includes updated AO. Their versatility will allow studies over much of the visible and IR portions of the spectrum.

A general statement may be appropriate here; since we often wish to measure properties of redshifted galaxies and AGN, it is natural for our new instruments to emphasize long-wavelengths, $\lambda \geq 1$ μm. For more local purposes, recall that observing long wavelengths allow one to see through foreground dust which would be many optical depths of visual light. One relatively new technique, near-IR AO, is a scheme that will clearly be of future importance in the studies of galaxies and QSOs even to $z \gtrsim 8$–10, a domain virtually unexplored so far.

6.2 New Techniques at the Focal Plane

Besides significant detector advances (Watson, 2004), new techniques at the focal plane may significantly increase the quality of data acquisition on Earth-bound telescopes. We discuss two advances, one applied primarily to imaging, and the other to spectroscopy. Both advances enable us to correct for atmospheric variability.

6.2.1 Adaptive Optics

Adaptive optics attempts to recover the theoretical angular resolution of a large telescope by correcting the negative, deleterious blurring of the Earth's atmosphere on the incoming stellar light wave-front. Ground-based large telescopes, especially those working in the visible region commonly only attain the theoretical resolution limits of ~20 cm diameter telescopes. The main reason for "astronomical seeing" are the temporal and spatial changes in the wave-fronts due to turbulence in our atmosphere; also image quality is lowered by semi-permanent telescope optics/mechanical errors, such as defocusing.

An active optics system, as the name suggests, works by means of continuous adjustments with small built-in corrective (mirror) optical elements that operate at temporal frequencies near 20 Hz. In practice this is far from easy, as the initially plane wavefront from a stellar source travels some 20 km through the "thick" atmosphere and then accumulates substantial phase errors (perhaps a few micrometers). These errors must be observed quickly and corrected every $\sim$2 millisecond! The field of view for optimally corrected distortions, the isoplanic angle, is typically very small; a few arc-seconds in visible light and a few tens of arc-seconds in the K-band (2.2 μm). In these circumstances (i.e., in the K-band, with a large telescope at a good site like Mauna Kea), the isoplanic angle is about $30''$.

To work rapidly, modern AO employs at least one small deformable mirror (typically size $D \simeq 10$ cm) located behind the focus of the telescope that views the (close-in) natural guide star, or an artificial laser "guide star", and makes rapid corrections in the collimated beam, just before the instrument detector. The basic layout of such an AO system is shown in Fig. 1.

The resulting improvements are in practice measured as the *Strehl ratio*, an image-quality index that is the ratio of the actual image peak intensity to that which would be obtained by the same telescope in the absence of turbulence (*i.e.*, only diffraction limited). A Strehl ratio ≥ 0.5 constitutes a notable resolution improvement. Table 1 shows Strehl ratios from Keck AO in three seeing ranges, and three wavelength ranges. Figure 2 shows Strehl ratios as a function of wavelength for various conditions, which produce different r.m.s. wavefront errors (Nelson & Mast , 2004). These values are characteristic of the image quality currently attained at the major innovative observatories.

TABLE 1: Strehl Ratios for Keck AO

Wavelength	Good seeing	Median seeing	Poor seeing
V 0.5 μm	0.06	0.01	0.001
I 1.0 μm	0.50	0.30	0.02
K 2.2 μm	0.87	0.78	0.43

Note: Strehl ratios from (Dekany , 1996).

What extragalactic cosmologic measures can we anticipate with near-IR AO in the next few years? Since AO shows the most dramatic improvement at long λ and with a large Strehl ratio (≥ 0.5), we can anticipate imaging at K-band with potential unresolved distant extragalactic samples – like QSOs. Are the $< 1''$ environments crowded at $z > 5$ (i.e., at scales $\lesssim 6$ kpc)? Do any appear to be close lensing pairs?

Since AO can be used with instruments of various utilities, we also consider narrow-slit AO IR spectroscopy as well as direct IR imaging. At Keck, the instru-

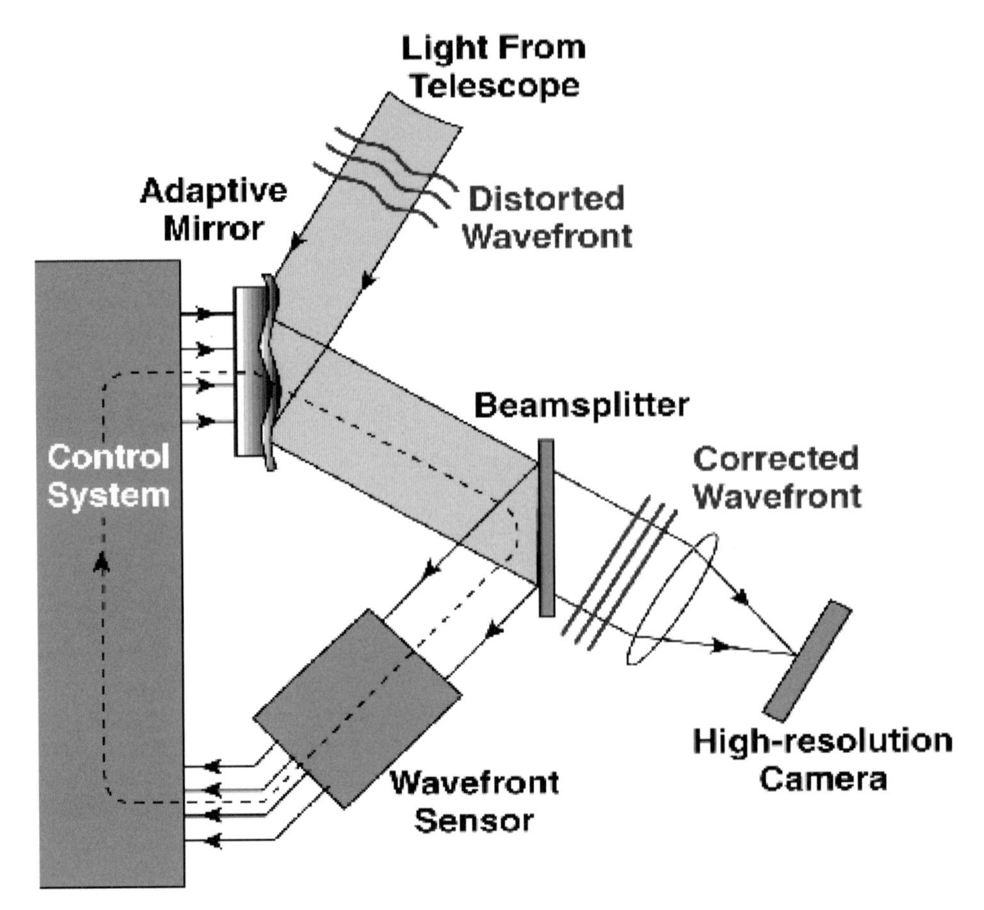

FIGURE 1: Diagram of an adaptive optical system. All AO systems work by first determining the shape of the distorted wavefront, then using an "adaptive" optical element (often a deformable mirror) to restore the wavefront to uniformity by applying an opposite distortion to the adaptive optical element. A point source of light (a bright star or an artificial star) is used as a reference beacon which enables wavefront analysis to be somewhat straightforward. Commands from the sensor activate the pistons that deform the mirror. In order to work well, the mirror's shape must be updated several hundred times a second. Image courtesy of Antonin Bouchez.

ment, OH-Suppressing InfraRed Integral-field Spectrograph (OSIRIS), is being commissioned to work in the AO-corrected field. The Integral Field Unit system (IFU, as it is called) is a fibre-fed spectrograph. It makes an array of spectra having a footprint comparable to that of a large galaxy at $z \sim 1$ uses a lenslet array to spectrographically sample rectangular patches of the sky at the high spatial resolution provided by AO, and with moderate spectral resolution ($R \sim 3800$). There are over 1000 of these lenslets, each capable of producing spectra in the z-, J-, H-, and K-bands. The light intercepted by the lenslets are then fed into optical fibres. This process avoids the usual ~40% loss to the fill factors of cir-

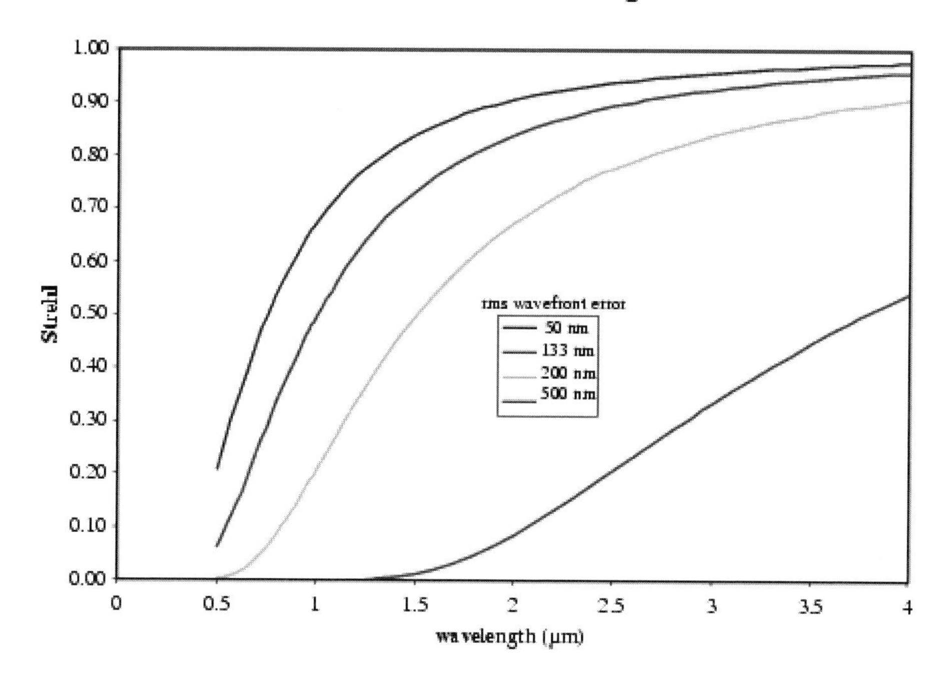

FIGURE 2: Strehl ratios as a function of wavelength of light, shown for various wavefront errors, given in micrometers. Graph from Nelson & Mast (2004). Smaller wavefront errors result in larger Strehl ratios. (Reproduced in colour in the colour section.)

cular fibres. Thus, the lenslet achieves a high fill factor, but the fibres increase sampling flexibility, optimizing their distribution to the spectrograph. Under spectral resolutions $R \sim 4000$, over 90% of pixels will be free of bright OH emission (Larkin et al., 2000). By avoiding spectral regions of OH emission, summed spectral channels can produce very deep near-infrared images. Because of the high spatial resolution, and the ability to remove sources of "background" emission, OSIRIS will reach point-source sensitivities 10 times that of the current Keck IR spectrograph. Lenslets may be of size $0''.02$, $0''.05$, or $0''.10$, leading to a field of view, or spatial resolution in the imaging mode of $1.''28 \times 0''.32$, $3''.20 \times 0''.80$, and $6''.40 \times 1''.60$, respectively. The Cambridge InfraRed Panoramic Survey Spectrograph (CIRPASS) is a similar IFU system to be installed on the 8-m Gemini South telescope (see Fig. 3). It can spectroscopically sample 499 locations with $0.25''$ fibres. The galaxy shown is at $z \simeq 1$, at which the proper physical scale of each fibre is approximately 2 kpc. Future AO units with resolution of about $0''.1$ on large telescopes will enable detailed analyses of galaxy kinematics and the star formation distribution.

Since IFU systems may have spatial resolution significantly under $1''$, this instrument may be useful for a wide range of astrophysical targets. Besides objects at high redshifts, and local AGN nuclei, IFU systems will be useful in analyzing surface details of solar system planets, moons and asteroids.

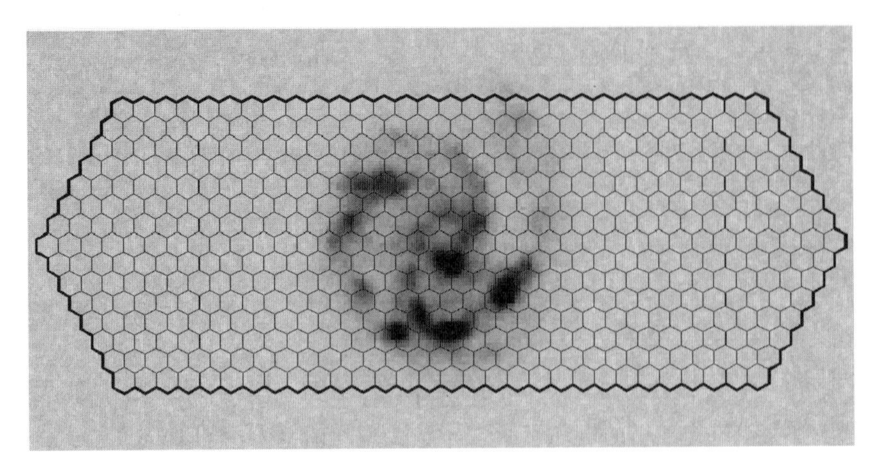

FIGURE 3: The CIRPASS IFU unit footprint for a B-band image of galaxy 4-550 in the HDF-North with a spectroscopic redshift $z \simeq 1.0$. The hexagonal pattern shows the placement of 499 lenslets and fibres through which spectra are read. Lenslets sample an angular diameter of $0''.25$; about 2.0 kpc in our favored flat Lambda cosmology. Image courtesy of A. Bunker and I. Parry (Institute of Astronomy, Cambridge).

If we assume Strehl ratios near 0.5 in K, the large telescopes of the future will be able to place $\approx$25 resolved pixel elements across distant galaxies ($z > 2$); this will be adequate to look for any systematic rotational support from nascent disks – spirals in the making – or, at another extreme, just a chaotic collection of sub-galactic clumps of star clusters at extremely early cosmic epochs. Here the "jumbled" structure would be well-suited to IFU spectroscopy, assuming strong spectral features were shifted into the observer's K-band (2.2 μm) as, for example, when [O III], 5007 $\longrightarrow$ K at $z \sim 3.4$. So far it doesn't seem practical to extend reasonable Strehl's to $\lambda < 0.9$ μm. And of course the well-corrected, isoplanic field size is still only $\sim 30''$ in the H-band ($\simeq 1.6$ μm).

As a "long-shot", it might even be possible to estimate dynamically the masses of the ubiquitous SMBH in early galaxy bulges at $z \sim 2$.

6.2.2 Nod and Shuffle: Optimized Background Subtraction

For years, extracting redshifts for galaxies in the "redshift desert" of $1 \leq z \leq 2$ has been frustrated by the lack of prominent emission lines in the observed optical bands when the spectrum is shifted by that degree. Recently, a new technique for background removal in spectroscopic observations on large, Earth-based telescopes has been developed, and may improve the quality of deep spectra by an order of magnitude. The cause for the problem in achieving optimal spectroscopy is that for faint objects, observations need to be of long duration. However, during periods in excess of a few minutes, sky lines can vary, leading to an imperfect subtraction.

Under this new technique, the telescope is "nodded" to a new position on the sky every 30 to 60 seconds. At the same time, the charges in the detector are shuffled*. This methodology uses a program to shuffle the charges a set number of pixels sideways between telescope exposures which have been "nodded" slightly between exposures. Charges on a CCD can be shuffled in unison a set number of pixels, moving the initial image over into a "storage" area. In practice, a given slitlet will use three different bands in the CCD, the original series of rows, and two equally wide "storage" bands on either side of the observing band. Following an exposure in the new position the telescope is nodded back to its original position. At the same time, the charges on the storage band are shuffled back onto the primary band, and the charges accumulated in the primary band are shuffled to the third band; the other storage area. This sequence is continued many times. The result will be a final primary image and a second storage image. The image difference of these two subtracts the sky, and leaves a positive and a negative image of the source. These will usually be combined by shifting the negative image the nodding distance and subtracting it from the primary.

This method is found to give an order of magnitude improvement in the quality of very deep spectroscopy, and has a most critical application in the near-IR ($\lambda \geq$ 7200 Å) where the telluric OH emission bands begin to gain in strength. At $\lambda > 8200$ Å the OH sky-lines can exceed the targeted galaxy continuum by a factor of 200–400, so that the precise sky-subtraction given with the nod and shuffle technique will help substantially in obtaining robust spectra of the faint, distant galaxies we discuss throughout the book.

6.3 New Tools to Reach Dim and Distant Galaxies

Here we emphasize some futuristic science goals in the study of evolving galaxies – and the instrumentation to take us there. Again, a common thread in our needs and their likely capabilities in the near future is the push toward high angular resolution observations at long wavelengths.

Among the ambitious plans for future telescopes, including those now being built, are optical telescopes and IR/mm arrays of enormous size and complexity. We discuss first the optical telescopes, then the radio region arrays.

6.3.1 Giant Optical Telescopes

Truly giant telescopes, significantly beyond the Keck telescopes and the VLTs, are possible only through the development of active optics, segmented mirrors, and AO. Segmentation and active optics have enabled large telescopes to be manufactured at modest cost by dispensing with the scale problems associated with giant mirror blanks and structural stability. These are now provided by actuators and a telescope feedback system. The AO is what enables these enormous planned

*Recall that the readout of a CCD occurs by the sequential shifting of pixel charges to a readout device at the edge of the CCD.,

telescopes to achieve resolutions approaching the diffraction limit,

$$\alpha = 1.22\frac{\lambda}{D}, \tag{1}$$

where the angle α is given in radians. Without the AO, resolution is limited to atmospheric seeing – only about $0''.5$ at best. The high resolution possible with AO on large "light buckets" will enable a detailed analysis of the structure of galaxies more than 7 Gyr old. It will probe the luminosity functions at $z \sim 4$ to well below L^*.

With these instruments, a 30-m class telescope should well be able to search many square arc-minutes in the near-IR J, H, and K wave-bands for galaxies that show no detectable visible light in I and z, just short of one micrometer. These would be I and z "drop-outs" at redshifts $z \geq 9$, a large jump toward the "boundaries" of our current limiting redshift. The concept has actually been under tests by an ESO group led Pelló et al. (2004). It is a difficult observational program employing the technology in use with our present 8-m class telescopes. Let us look at these new designs, working our way from the smaller to the largest planned optical telescopes.

The Giant Magellan Telescope, would feature seven 8.4 m mirrors, equivalent to a 21-m diameter telescope. One telescope is planned for the plains of northern Chile, and another may be planned for Antarctica, where conditions are optimal for near-IR observations with or without AO. It is being discussed by a consortium including Carnegie Observatory, Harvard University, the Smithsonian Astrophysical Observatory, the Massachusetts Institute of Technology, the University of Arizona, and the University of Michigan.

The Thirty Meter Telescope (TMT) is a planned 30-m telescope being developed by the National Optical Astronomy Observatory and other major universities such as the University of California and Caltech. With AO on, the isoplanic field will be $\sim 52''$. It is capable of delivering Strehl ratios of ~ 0.5 at wavelengths $\lambda \gtrsim 1$ μm, providing resolution of 10 milliarcseconds. It is expected to operate in the $\lambda \sim 1$–30 μm range. This telescope should have a resolution comparable to the Atacama Large Millimeter Array in Chile (ALMA, see below), the planned sub-millimeter array, and should be complementary for the study of star-forming regions and other uses. Multiplexed spectroscopy should enable the acquisition of thousands of galaxy spectra simultaneously. The planners anticipate that the TMT will play the same role to the planned James Webb Space Telescope (see below) that Keck does for the HST. The consortium hopes to complete the project within a decade, though it is still far from a realistic start.

Finally, there is the OverWhelmingly Large telescope (OWL), proposed by the European Southern Observatory (ESO – see Fig. 4). The primary rationalé for such a large telescope is to search for the "holy grail" of today's astronomy – extra-solar Earth-like planets orbiting the nearest stars. This task requires a light-gathering power equivalent to a $\sim$80-m diameter mirror. Early planning is now under way; designers expect to achieve 2 arcmin of diffraction-limited field of view in the visible. The primary mirror would have perhaps 3000 segments of

~1.6 m each. The secondary will be a flat segmented mirror of 25.6 m diameter. There are four corrector mirrors, the first two of which are active. The structure has a moving weight of 14,200 tons! The field of view would be 10 arc minu (seeing-limited), and have a 2 arcmin isoplanic field. The telescope, with AO on, is capable of working near the diffraction limit of ~1 milliarcsecond, and may probe to amazingly faint visual magnitudes near $V = 37$ mag. A benefit of this profound leap in resolution and light-gathering power will be the distance determinations of the primary distance indicators such as Cepheids, Novae, and others, to $z \sim 1$, where the difference between cosmological models is more pronounced. Type 1a SNe would be detectable to $z \sim 5$. It will be useful in sampling the epoch of re-ionization by Pop III stars. For instance, at 1 milliarcsecond resolution, and with a flat $\Lambda = \mathrm{h} = 0.7$, scales of less than ~6 pc at $z = 6.5$ could be probed.

FIGURE 4: Artist's conception of the giant OWL telescope layout. See http : www.eso.orgprojectsowl

Of course, all of extragalactic science on the frontier in a decade would benefit from a well-engineered OWL.

6.3.2 Allen Telescope Array

The use of modern large and long-wave interferometers at sub-millimeter, millimeter, and radio (meter) wavelengths will have immediate application in several extragalactic sub-fields. A 21-cm line-survey is planned with the (under construction) Allen Telescope Array (ATA) in California. With this instrument, one can imagine a "blind search" for local neutral H gas clumps, perhaps not even associated with well-evolved galaxies of dark matter and stars. The search could be independent of galaxy clustering; as envisioned with the ATA, the sky coverage for H I will be 17 times the area possible with the current VLA and also offers excellent frequency coverage (from ~500 MHz to 11.2 GHz), so its 21-cm (1420 MHz) search can supplement optical surveys of nearby gas-rich galaxies, and possibly detect low surface brightness galaxies, and high-velocity H I clouds in the periphery of nearby galaxies beyond the Local Group. Each resolution element of the ATA will reach a 5-σ sensitivity of $\sim 6 \times 10^3\,\mathrm{M}_\odot\ \mathrm{Mpc}^{-2}$ in a single four hour integration with an impressive velocity resolution of 20 $\mathrm{km\,s}^{-1}$. Such a survey would have a mass sensitivity of about $10^6\ \mathrm{M}_\odot$ for a 21-cm linewidth of ~30 $\mathrm{km\,s}^{-1}$at a mean recessional velocity of 1000 $\mathrm{km\,s}^{-1}$ (similar to a Virgo cluster distance of ~ 15 Mpc). It will also be possible to extend an H I-survey to around $z \simeq 0.15$ for L_* (~MW mass) systems, and occasional very massive H I galaxies might be well matched to the Sloan Digital Sky Survey (of course emphasizing the stellar component) to $z \simeq 0.2$.

6.3.3 The Square Kilometre Array

The Square Kilometre Array (SKA) is a proposal to build a interferometric radio telescope with a total antenna collecting area of one square kilometer. The primary incentive for building this telescope is to observe and map structures, and substructures forming in the early Universe at redshifts $z \leq 20$. It will be capable of detecting H I clustered around galaxies at cosmological redshifts. In addition, the angular resolution may be about a tenth of an arcsecond, good enough to resolve galaxies. This resolution would enable the study of stellar activity and magnetic fields by the detection of radio synchrotron and free-free continua in evolving galaxies. Because of its high angular resolution, the SKA will be able to measure galaxy rotation curves to $z \simeq 0.5$.

The debate over the best location for the SKA telescope is still under discussion at the time of writing. However, Western Australia is thought to be the leading contender, with South Africa also possible. The telescope is planned to have a *central site*, an area of about 100 km diameter, and a number of "remote" stations. Depending on the size of the individual antennas, the telescope could have tens of thousands of antennas. For instance, if each antenna is to be 12 meters in diameter, there would have to be 8,842 antennas. While more dishes would make for higher spatial resolution, the cost of the correlator, and the optical fibre required to hook them together, would increase dramatically with the number of antennas.

Clearly this is a gargantuan effort. The proposed time table suggests that planning could be completed, and construction could begin by 2010, and perhaps it could begin operation by 2015.

6.3.4 Atacama Large Millimeter Array

Another scientific question where technology improvements will likely be valuable is the time scale for the build-up of "heavy elements" like C and O in young galaxies; are they built to near solar abundance in an initial starburst early in the galaxy's formative years? We'll need the Atacama Large Millimeter Array

(ALMA) to break ground on that topic; the astrophysical arena of ALMA is the realm of molecular emissions from low-lying rotational levels, continua from cold dust, and some redshifted atomic lines. Millimeter molecular lines of CO especially are starting to play an important role in galaxy/QSO astrophysics at large redshifts. This lead will be amplified by the power of the ALMA array compared to the present capabilities of pioneering instruments like SCUBA (the Submillimeter Common-User Bolometer Array), commonly used on the 15-m JCMT at $\lambda = 450$ μm and $\lambda = 850$ μm.

ALMA is probably going to be our most important new instrument to be constructed and used on a reasonable time-scale. It is a project of such potential import that it has been made very "multi-national" and "multi-continental". The costs are being divided between North American partners (NSAF/NRAO+NRC Canada), and Europe (ESO + Spain), with Chile providing the high and dry observing site. Japan may soon join this project. The array should be operational by 2011. The nations are building toward a powerful interferometer with a total of 64 12-m sub-millimeter antennae of high quality. It will operate in four frequency bands, covering the main atmospheric windows between 4 mm and 0.4 mm. The present pair of SCUBA bands, at 450 μm and 850 μm, will be two of these windows.

ALMA will be particularly important in answering questions on the early evolution of dusty and often gas-rich young galaxies. Following Combes (2001) we present three arguments that became drivers for ALMA. First, we now know that there is a factor of $\sim$10 increase in the SFR in normal galaxies from $z = 0$ to $z = 1$, so we may want to learn more of the evolution of gas-rich systems near unit-z. Secondly the most active or violent starbursts seem to be the most obscured by dust – we observe the re-radiated UV energy in the infrared. Finally, detections in the millimeter and sub-millimeter regions are favored owing to the emission SED of dust emission in this part of the spectrum. For, as one looks to higher redshifts, the loss in flux due to distance is compensated by the increased flux in sampling sampling wavelengths closer to the maximum emission at around a rest $\lambda \sim 100$ μm. And in some cases, we hope this large negative K-correction will permit the discovery of dusty galaxies out to $z \sim 10$.

The main current problem with present-day millimeter surveys is the specific identification of the millimeter sources; much of the problem relates to the low spatial resolution of instruments like SCUBA – with a resolution of approximately $15''$. With ALMA, the angular resolution will exceed $0''.2$ for a medium-length configuration spacing!

The ALMA detections will also include much line radiation from molecular rotational transitions besides the redshifted light from cool dust. The elements contributing to the molecules we shall observe are formed from mainly H, C, O and N. The CO molecule is, of course, the main tracer of molecular hydrogen gas, especially in our Milky Way galaxy. Velocity fields will be well-studied and in some cases, the millimeter domain will open up studies of different objects than are seen in the optical region due to dust extinction. The redshifts of these "molecule factories" can be measured by their CO lines; at high redshifts, fine structure lines of ions like C^+ (at $\lambda_0 \simeq 150$ μm) will be observable.

The survey sensitivity of ALMA will be impressive; a sub-millimeter search for distant starburst galaxies should not be limited by source confusion, as is currently the case with SCUBA. A $4' \times 4'$ field could be observed to a sensitivity of 0.1 mJy (that is 20 times lower than SCUBA surveys now) in 2 weeks. Over 100 galaxies should be detected and for many with CO lines detected, their redshifts will come with unusual ease, perhaps up to $z = 10$! Thus, dusty starbursts will now show an "open door" for astronomers as opposed to the dusty and obscured window now facing these tracers of heavy-element building.

Baker's work on cB58 (Baker et al., 2004) – a lensed LBG system – simulates the heightened depth and spatial resolution that ALMA will provide unlensed high-redshift objects. These authors report the detection of CO(3-2) emission, and Hα, with line-widths that may provide information about the velocity dispersion of LBGs, and provide a self-consistent picture of past star formation.

6.3.5 The Spitzer Space Telescope (SIRTF)

Formerly named the Space Infrared Telescope Facility (SIRTF), the Spitzer Space Telescope represents the final mission in NASA's Great Observatories Program (the others being HST, the Compton Gamma Ray Observatory, and the Chandra X-Ray Observatory). It is a space-borne, cryogenically cooled observatory capable of observing (imaging and photometry) in the range $3 \leq \lambda \leq 180$ μm. Spectroscopy can be performed in the range 5 to 40 μm, and spectrophotometry from 50 to 100 μm.

The telescope is in a novel orbit, designed to trail the Earth at an ever-increasing distance, increasing by about 0.1 AU per year, thereby progressively removing it from the thermal effects of the warm Earth (it has a shield for the Sun). This orbit will also keep the telescope away from the radiation belts that surround our planet.

The 85-cm diameter mirror is made of lightweight beryllium, and cooled to less than 5.5 K. The cryogenic supply is expected to last at least 2.5 years, but it is hoped that it can be stretched to 5 years of operation.

Thermal radiation from a black body of temperature T (in kelvins) has a wavelength of maximum flux $\lambda_{\rm max} \simeq 3 \times 10^3/T$ μm. For $\lambda = 100$ μm, thermal radiation of temperature 10 K would be prominent. This is of the same order of temperature as cool interstellar dust. In addition, many molecules, including organics, have spectral lines in the range detectable by Spitzer.

There are three Spitzer focal plane instruments, all designed to investigate the major science objectives of this telescope (Werner et al., 2004). The instrument most relevant to galactic thermal emission is the Infrared Array Camera (IRAC) (Fazio et al., 2004). In particular, it is designed to study the evolution of normal galaxies to $z > 3$ with deep, large-area surveys; a four-channel camera provides simultaneous imaging in 3.6, 4.6, 5.8, and 8 μm. Each of the four detector arrays are 256×256 pixels in size, with a pixel size of $1''.2$ on the sky. The major science objectives of IRAC that are relevant to extragalactic studies are to study the IR output of young galaxies in the early Universe, and to study ultraluminous IR galaxies and galactic nuclei.

The Infrared Spectrograph (IRS) onboard the Spitzer Space Telescope (Houck et al., 2004) is composed of four separate modules, two with low resolution spectroscopy ($R \sim 90$) from ~4 to 26 μm and ~14 to 40 μm, and two with high resolution ($R \sim 600$) spectroscopy for ~10 to 20 μm and for 20 to 40 μm. The low-resolution modules use a long slit in the focal plane, and the high-resolution modules use a cross-dispersed echelle design. This instrument can be used to monitor the details of dust emission in the 10 to 30 μm range, and help us to understand the environmental dependencies of extinction.

The Multiband Infrared Photometer for Spitzer (MIPS) rounds out the instrument payload, concentrating on the far infrared. It provides imaging bands at 24, 70, and 160 μm (with bandwidths of 5, 19, and 35 μm, respectively), and low-resolution SEDs between 52 and 100 μm. Detectors are arsenic-doped silicon, gallium-doped germanium, and stressed gallium-doped germanium, respectively. At 24 μm, photometric errors are about 1%, while for the other arrays, the best that can be hoped for is about 10% accuracy. The arrays contain 128×128 pixels at 24 μm, 32×32 at 70 μm, and 2×20 pixels at 160 μm (Rieke et al., 2004). The small size of the mirror limits the angular resolution to $6''$, $18''$, and $40''$, respectively. These instruments will be important in analyzing the cool dust component, and to bridge the gap between the NIR data and the Earth-based sub-millimeter bolometer arrays.

In addition, Spitzer is dedicated to a series of "heritage" surveys. SWIRE, the SIRTF Wide-Area Infrared Extragalactic Survey, will use IRAC and MIPS, to survey high latitude fields totaling 60–65 deg^2. This imaging survey is to trace the evolution of dusty, star-forming galaxies, evolved stellar populations, and AGN, as a function of environment for $z \lesssim 3$ (Lonsdale et al., 2003). For studies of galaxy formation, this will be invaluable since long-wave photons are significantly less extincted than the visible. In addition, angular correlation functions will be derived for each of the IRAC channels. SWIRE will also be used to attempt identification of any X-ray selected clusters ($z > 0.5$) that are not detected optically (Donahue et al., 2002; Lonsdale et al., 2003).

For comparison, we note that the SIRTF Nearby Galaxy Survey (SINGS) is a comprehensive infrared imaging and spectroscopic survey of 75 nearby galaxies. It is hoped that this analysis will enable the development of improved diagnostics of SFRs in galaxies by comparison with UV, Hα and IR data (Kennicutt et al., 2003).

6.3.6 The James Webb Space Telescope

The James Webb Space Telescope (JWST) will have a primary diameter of about $D \approx 6.5$ meters, and will be sent to the Earth's L2 Lagrangian point, a stable orbit farther than that of the Moon, but on the night side of the Earth relative to the Sun. It is scheduled to launch in 2011. The emphasis in sensitivity is to the IR in deference to the search for high-redshift objects. It is to be optimized from $\lambda = 1$ μm to 5 μm in order to concentrate on the redshifted rest-optical and UV radiation from the time interval when galaxies began to form – well after recombination and the source of the cosmic microwave background, but before the "recent" past ($\sim$ 10 Gyr ago).

We would like to know much about the first sources of light that re-ionized the previously dark Universe. Were they large stars powered by thermonuclear fusion, or massive objects accreting infalling matter? JWST spectrometers should have adequate sensitivity to settle this problem.

Another matter of cosmological import is the angular diameter of compact objects. For a flat Λ model, the angular diameter distance attains a maximum at about $z = 1.5$, with $d_{\mathrm{A}} = 1753\ \mathrm{h}_{70}^{-1}$ Mpc, while at $z = 10$, $d_{\mathrm{A}} = 862\ \mathrm{h}_{70}^{-1}$ Mpc. Thus, the same object, seen alternatively at $z = 1.5$ and at $z = 10$ will have an angular diameter more than twice as large at $z = 10$.

To explore some of these questions the JWST 6.5-m telescope will be required to demonstrate the following characteristics:

- Excellent sensitivity over the 1–5 μm region of the spectrum, with 10 times the light-gathering power of HST, and new solid-state devices with high photon efficiencies.

- Diffraction limited imaging at 2 μm. In space, no AO is needed to approach the theoretical capabilities of the telescope.

- The capability of multi-object spectroscopy in the $1.0 \lesssim \lambda \lesssim 5.5\,\mu$m.

- good sensitivity in the mid-IR (5–20 μm), with single-slit spectroscopy at those longer wavelengths.

Efficient photon detection in the IR requires technical advances over the standard CCDs, which is essentially transparent to IR photons. However, fairly satisfactory substitutes are available. At this early stage, the following detectors should be taken as merely illustrative. An indium antimonide substrate (InSb) may form the basis for an efficient detector in the near-IR. It has a high quantum efficiency at wavelengths $1.0 < \lambda < 5.5\ \mu$m. In the mid-IR, arsenic-doped silicon detectors (Si:As) can be used over wavelengths $5 < \lambda < 28\ \mu$m. The near-IR camera could use a mosaic of 16 1K$\times$1K detectors, while the mid-IR will probably have just one.

The field of view of the near-IR camera is large; $10' \times 10'$, while that of the MIR camera is more modest, $2'$ square.

6.4 Summary

The technology available to the astrophysicist will continue to expand almost geometrically – telescopes will continue to get bigger, and better. We have seen that realistic plans for AO-assisted telescopes with up to 100 times larger light gathering power than the Keck telescopes are being actively pursued. Integral field spectroscopy on these telescopes would bring a dynamical analysis of high-redshift objects within the range of feasibility. A parallel development with millimeter and sub-millimeter arrays is also taking place. The next-generation radio telescope ($\lambda \gtrsim 1$ cm), SKA, will extend the utility of centimeter-wave astronomy to cosmological issues through its sensitivity to 21-cm hydrogen emission at a spatial resolution of ~ 0.1 arcsec. The millimeter, sub-millimeter and entimeterm wave telescopes will provide spatial resolution comparable to that of the planned giant optical telescopes. Finally, two space telescopes, one, Spitzer, already in service, and the other, JWST, will ultimately provide great advances in the IR and NIR. These telescopes will complement the Earth-based telescopes in the same way that HST and Keck are complementary – space telescopes for very deep, precise work, and large ground telescopes as the work-horses of analysis.

REFERENCES

Baker, A. J., et al., 2004, ApJ, 604, 125

Combes, F. 2001, "Galaxy Evolution with Alma" in *EDPS Conference Series in Astronomy and Astrophysics*, astro-ph/0107403

Dekany, R., "The Palomar Adaptive Optics System", in *OSA 1996 Adaptive Optics Topical Meeting Proceedings*, p. 40, Optical Society of America, (1996).

Donahue, M., et al., 2002, ApJ, 569, 689

Fazio, G. G., et al., 2004, ApJS, 154, 10

Houck, J. R., et al., 2004, ApJS, 154, 18

Kennicutt, R. C., et al., 2003, PASP, 115, 928

Larkin, J. E., et al., 2000, in ASP Conf. Ser. 195: *Imaging the Universe in Three Dimensions*, 508

Lonsdale, C. J., et al., 2003, PASP, 115, 8

Nelson, J. & Mast, T. 2004, *Conceptual Design for a 30-Meter Telescope*, in CELT Report Number 34

Pelló, R., et al., 2004, AAP, 416, L35

Rieke, G. H., and colleagues 2004, ApJS, 154, 25

Watson, F., 2004 in *Astrophysics Update*, ed. J. W. Mason, Heidelberg: Springer-Praxis, p. 181

Werner, M. W., et al., 2004, ApJS, 154, 1

Chapter 7

Briefly: Some Overall Conclusions and Problems

7.1 General

Galaxies can now be observed probably forming and certainly evolving over the large span of cosmic time implied by their range of redshift, $2 \lesssim z \lesssim 6$. Robust conclusions to longstanding astrophysical queries about the "early life" of galaxies do not come quickly or even with confidence. However, there are some strong implications for our field, based, as usual, on incomplete observations and theory which may have an unwarranted popularity. Nevertheless, it seems clear that isophotal sizes of galaxies at high redshift are significantly smaller than earlier expectations. As shown in Fig. 4 Chapter 3, mean half-light radii approach 1 kpc at $z \simeq 5$. A good example is the pair of $z = 5.34$ galaxies shown in Fig. 1 that are detected (and not detected) in F814W (and F606W) HST bands. The deconvolved half-width, half-maxima of the images are $0''.24$ and $0''.12$, for the larger and smaller, respectively. For a flat $\Lambda = 0.7$ cosmology, this refers to 1.4 h_{70}^{-1} and 0.7 h_{70}^{-1} kpc, respectively. As we saw, modeling these small systems by the collapse of a homogeneous Jeans mass cloud (Chapter 3, §4) does not seem to fit observations well since larger diameters are predicted. The inside-out galaxy formation model from a concentrated and inhomogeneous cloud produces smaller stellar condensations that are consistent with observations.

7.2 Galaxy Growth

Some concepts of galaxy youth and galaxy evolution are recurring. To this observer they include support for the growth of small-mass halos with smaller mass baryonic galaxies within. Eventually the dense (or most-dense) of these structures attain Milky Way dimensions. Most of this hierarchical scheme is carried out by modest to major mergers. Often these events or their past are detectable through

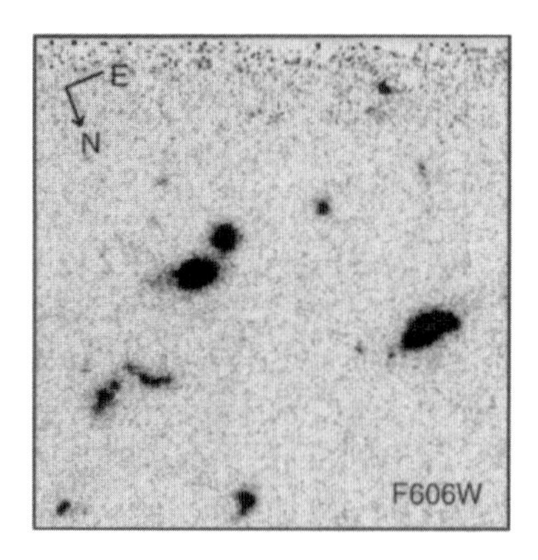

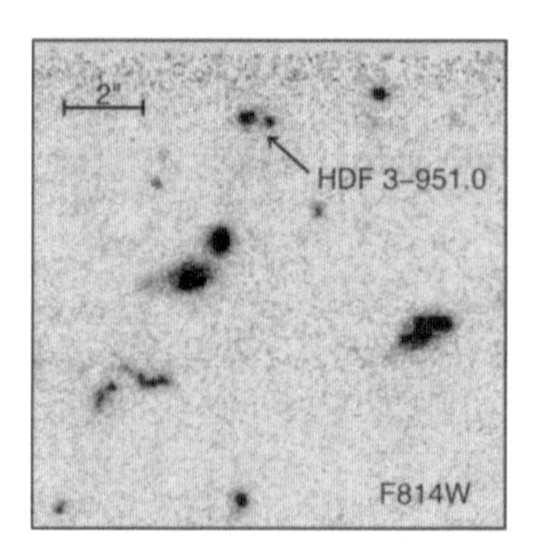

FIGURE 1: HST images of a close pair $z = 5.34$ systems, HDF 3-951.0. Deconvolved half-width, half-maxima of images are $0''.24$ and $0''.12$, for the larger and smaller components, respectively. For a flat $\Lambda = 0.7$ cosmology, this refers to 1.4 h_{70}^{-1} and 0.7 h_{70}^{-1} kpc, respectively. Also note the weakness of the images in the V-band (F606W); this is due to the IGM absorption below Lyα in the system rest-frame. Image from Spinrad et al. (1998)

the resultant gravitational "mess" being detectable by an atypical post-merger structural character (*cf.* Conselice et al. 2003).

However, substantial mass-infall from under- or un-luminous sub-galactic perturbations or mini-galaxies, perhaps of Jeans–scale, is also plausible. The intensity and frequency of the merging process declines with time. The E galaxy end-products are now regular and have been astrophysically "inactive" over at least the last ~1/2 Hubble time ($z \lesssim 0.7$). This quick evolution or perhaps "monolithic evolution" was clearly truncated 4–7 Gyr ago – few young stars are found to play a role in (especially) cluster E and S0 galaxies. An older stellar population without current SF lives there. The mild stellar evolution of the E galaxy population over $0.0 \leq z \leq 1.5$, at least for massive (luminous) ellipticals has been studied by (Cimatti et al., 2004), and by Drory et al. (2004), and by Bell et al. (2004). All of these authors conclude that the last epoch of substantial E-galaxy SF was long ago, at $z > 2$. The relative fraction of an older population increases with time. However, small SF events may happen even for these "rather dead" galaxies and large spirals. In our MW system, a modest accretion of gas and/or small satellites continues to the current epoch. These are definitely minor mergers, and at least several are needed to impact the overall galaxy mass, and actually lead to a noticeable morphological change.

7.3 Proper Sizes

A strong direct confirmation, perhaps still with some biases, rests on recent high angular resolution images of the distant Universe (the GOODS fields and other selected areas for deep surveys). Publication of typical angular sizes yielded galaxy radii as a function of their (photo z) redshifts. Figure 5 in Chapter 3 illustrates the mean galaxy metric (proper) radii as a function of redshift, combining the data of Ferguson et al. (2004) and Bouwens et al. (2004). The trend of increasing scale with decreasing redshift is obvious: at the mean survey redshift of $z \simeq 3$ $r_{1/2}$ is only 2 kpc, while at low redshifts, values approach 5 kpc. So the early galaxies ($z > 3$) would appear to truly be pre-galactic star (+gas) clumps, in the sense of a linear dimension trend. The surveys mentioned above found essentially *no* large disks at high redshifts.

7.4 SFR and Mergers

SFRs are measurable for galaxies bright enough to enable spectroscopy or robust photo-zs. It may be the case that even minor-mass mergers stimulate the galactic SFR for modest periods of time. Thus, galaxies that are "disturbed-looking" might be relatively bright in their emitted ultraviolet ($\lambda_0 \leq 3000$ Å).

Unfortunately, we probably do not detect small galaxies with a low SFR. Their ultraviolet fluxes are correspondingly weak, and that implies an awkward selection effect may be in place at $z \geq 3$ when the emitted UV spectrum moves out to the optical domain of the observer's frame.

We would like a volume-limited sample in which to pursue this correlation between morphological signs of mergers and the (then) current SFR. Is there really a positive correlation between SF and merging (as measured by an asymmetry index)? The global evidence for large shifts in the SFR alone, from photometry and spectrophotometry indicates a broad first SF peak as one approaches $z \simeq 1.5$–2.0. This is shown in Fig. 2 (Steidel et al., 1999). The difference between $z = 0$ (local) and $z \simeq 1.0$ is close to an order of magnitude!

7.5 The Re-ionization by Stars in Young Galaxies

Controversy still swirls about the last and latest ionization of the IGM, and its median redshift, perhaps between $z = 6.5$ and $z \simeq 20$. It is the surge of UV ionizing photons that transforms a dark time interlude to a IGM situation that is ionized. The degree of IGM ionization is almost complete in quick order; for $z < 4$ only ~ 1 part in 10^5 is neutral H.

To accomplish the rapid re-ionization (between $z \sim 20$ and $z = 6$), we require a plentiful supply of photons from some UV-emitting sources. This turns out to be a substantial contemporary problem. We cannot easily appraise the overall census of UV-emitters at $z > 5$. QSOs and fainter AGN (even X-ray sources of moderate luminosity) are collectively rather rare at $z > 5$. Certainly few very distant QSOs have been located by the full spectrum of modern surveys at many disparate wavelengths, and are unable to account for the ionizing photons (Barger et al., 2003). Failing AGN, we fall back on individual massive, probably metal-poor OB stars in systems of the size of LBGs or smaller for ionization of the IGM.

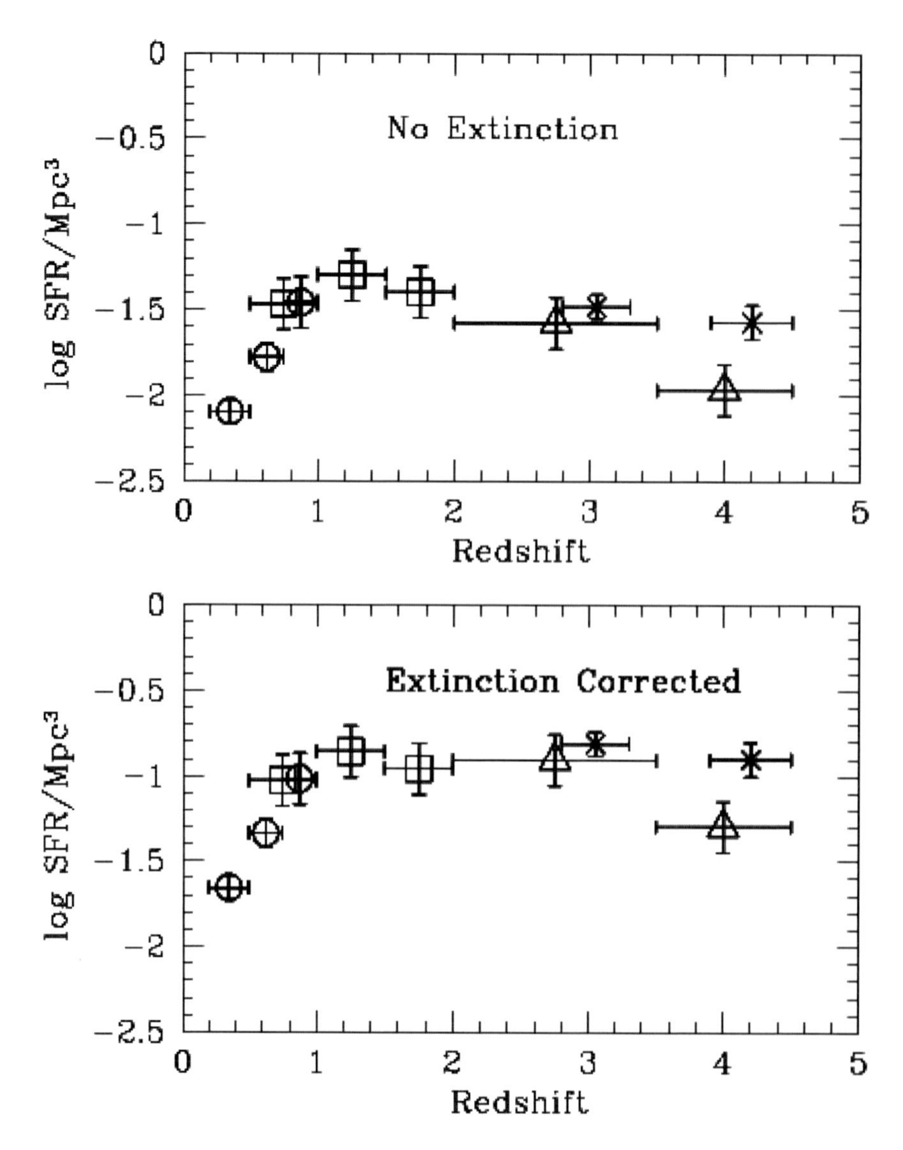

FIGURE 2: The uncorrected (top) and extinction-corrected (bottom) SFR density ("Madau diagrams") as a function of redshift, where the SFR is given in solar masses per year. Different symbols refer to data from different sources – see Steidel et al. (1999). Though the plot assumes an Einstein–de Sitter cosmology with $H_0 = 50\ \mathrm{km\,s^{-1}\,Mpc^{-1}}$, the general shape is maintained, even in a flat λ universe.

Lehnert & Bremer (2003) showed that LBGs do not supply enough ionizing UV to keep the IGM ionized (see chapter 4, §4.23). But this question may also be asked of low luminosity ($< \mathcal{L}^*$) UV-emitting galaxies. To assess the ability of small star forming galaxies to re-ionize the universe, we must know the UV escape fraction, f_{esc}. Current parameterizations of f_{esc} are of two forms; one simply uses the fraction of 900 A photons (just below the Lyman limit) that escape from a galaxy, n_{esc}/n_{emit} (Leitherer et al., 1995), where n_{emit} is a result of model galaxy studies. The other method normalizes this fraction to the fraction of 1500 Å photons that escape the galaxy (Steidel et al., 2001). Because only about 15% to 20% of λ1500 Åphotons escape a star forming galaxy, the Steidel et al. escape fraction ($f_{esc} \gtrsim 0.5$ for $z \sim 3.4$ galaxies) is about a factor of 5 larger than that of the Leitherer et al. value, which was based on local compact starburst galaxies. We use the Leitherer et al. parameterization here.

With escape fractions $f_{esc} \geq 10\%$, small galaxies may just be able to maintain IGM ionization. This escape fraction is controversial and quite uncertain; circum-galactic or circumnuclear H is very opaque below $\lambda_0 = 1216$ Å, and at $\lambda_0 912$ Å and below. On the other hand, the faint-end slope of the UV luminosity function is steep ($\alpha \simeq -1.6$; Steidel et al. 1999), so there may be very a high density of such star formers. On balance, the faint galaxy theory for re-ionization cannot yet be deemed robust, though the apparent lack of another candidate augurs for a closer look.

Yan & Windhorst (2004) constructed a model of the luminosity function of UV-emitting galaxies. To make the IGM completely ionized at redshift $z \sim 6$, Yan et al. demand a slightly steeper-than-routine faint-end slope of the galactic luminosity function of $\alpha = -1.6$ to $\alpha = -2.0$. Perhaps galaxies will, as individuals, directly give evidence for escaping flux below λ_0 912 Å. But the problem might have to wait until spectroscopic observations with the futuristic large telescopes (see Chapter 6 in this book) become routine.

7.6 The Re-Ionization State of the Early IGM

The Gunn–Peterson (G-P) effect in QSO spectra is our main guide to the beginning of the re-ionization era. We think the G-P troughs are crude measures of the scales of still partly neutral H gas in the densest filaments occupied by the existing and collapsing structures, baryonic and dark matter.

With difficulty our study of the G-P troughs in $z > 6$ QSO spectra allow us to measure the relative amounts of remaining neutral gas. We observe below the Lyα and Lyβ lines and extract optical depths and relative transmissions in the almost dark spectral troughs (Djorgovski et al., 2001; Becker et al., 2001; White et al., 2003). The most recent spectroscopic interpretations of the change in G-P depth as a function of redshift (White et al., 2003; Songaila, 2004) appear to show a transition from a gradual increase in the neutral H absorption and its optical depth from $z = 5$ to $z \sim 5.7$, and a sudden increase in optical depth beyond that (see, Fig. 4 in Chapter 4). On the other hand, WMAP data suggests that the Thompson re-ionization optical depth is $\tau \simeq 0.17 \pm 0.04$ (68%) (Kogut et al., 2003), implying a redshift of re-ionization $z \sim 17$ (Spergel et al., 2003).

This larger-than-expected redshift value (given the observations of the G-P trough) suggests a possible problem. Recently, however, Gnedin (2004) presented results of cosmological modeling which is consistent both with the WMAP polarization data and a re-ionization redshift of $z_{\rm re} = 6.1$. Some of these scenarios for re-ionization are shown in Fig. 3. Gnedin explains this by pointing out that at $z \sim 6$, a neutral fraction of 2×10^{-4} can effectively remove all Lyα photons, even if the gas is under-dense (relative to the average) by a factor of 10. A final stage of re-ionization, to neutral fractions less than 1 part in 10^5, is required before the Gunn–Peterson troughs disappear. This implies perhaps a long and interesting re-ionization of the Universe.

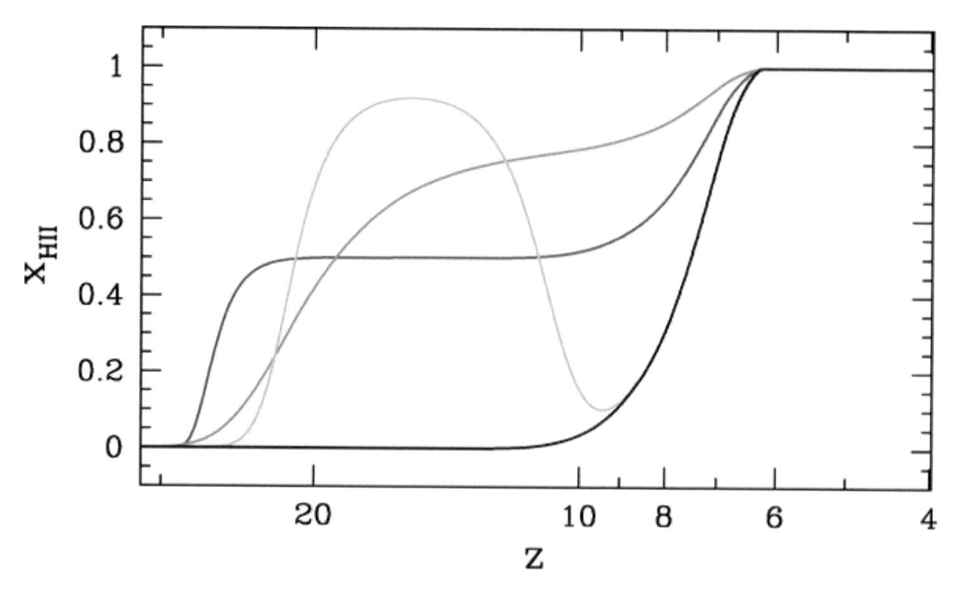

FIGURE 3: Alternative histories for re-ionization of the Universe that are consistent with WMAP findings and the observed Gunn–Peterson troughs of $z \sim 6$ QSOs. The ordinate X_{HII} is the fraction of hydrogen which is ionized. Illustration from Gnedin (2004).

The details we may extract with the G-P observations must, of course, be consistent with any physical conclusions about the sources of ionization (see §7.5). Unfortunately, those Pop III stars, as many call them, have proven to be elusive; a few very metal poor and low-mass galactic halo stars survive. There are no surviving large-mass extremely metal-poor galaxies. Perhaps we have come on the scene a bit late... Indirectly we still imagine the largest Population III stars (~ 200 M$_\odot$?) might be required to do much local re-ionization and account for the polarization of the microwave background measured by WMAP.

7.7 Evolution of QSOs

Many now believe that QSOs are not major re-ionizers of the IGM (e.g. Steidel et al., 1999; Bremer et al., 2004; Barger et al., 2003; Lehnert & Bremer, 2003; Boyle et al., 2000). Schmidt et al. (1988); Schneider et al. (1989); Schmidt et al. (1991)

showed that the comoving space density of bright QSOs increases dramatically back to $z \sim 1.5$, and declines dramatically beyond $z \sim 3.0$. Because some AGN may be obscured by circumnuclear dust, they may not be detectable with optical surveys. However, hard X-rays can penetrate the dust torus. Chandra X-ray data with deep optical searches was used to show that bright, QSOs ($44 \lesssim \log L_X \lesssim 45$ ($\mathrm{erg\,s^{-1}}$)) have a nearly constant comoving number density between $z = 1.5$ and 3.0. The comoving density of fainter QSOs ($43 \lesssim \log L_X \lesssim 44$) increases gradually with time over the interval $6 \gtrsim z \gtrsim 0.1$ (Barger et al., 2003), as shown in Fig. 4 (solid circular points). Steffen et al. (2003) found that the comoving density of bright ($L_X \geq 10^{44}\,\mathrm{erg\,s^{-1}}$) broad-line QSOs declines with decreasing redshift at $z \lesssim 1$, though the number of broad-line QSOs of all luminosities holds relatively constant, suggesting strong luminosity evolution. Narrow-line AGN (e.g., Seyfert 2s) dominate the XLF at low-z (Steffen et al., 2003).

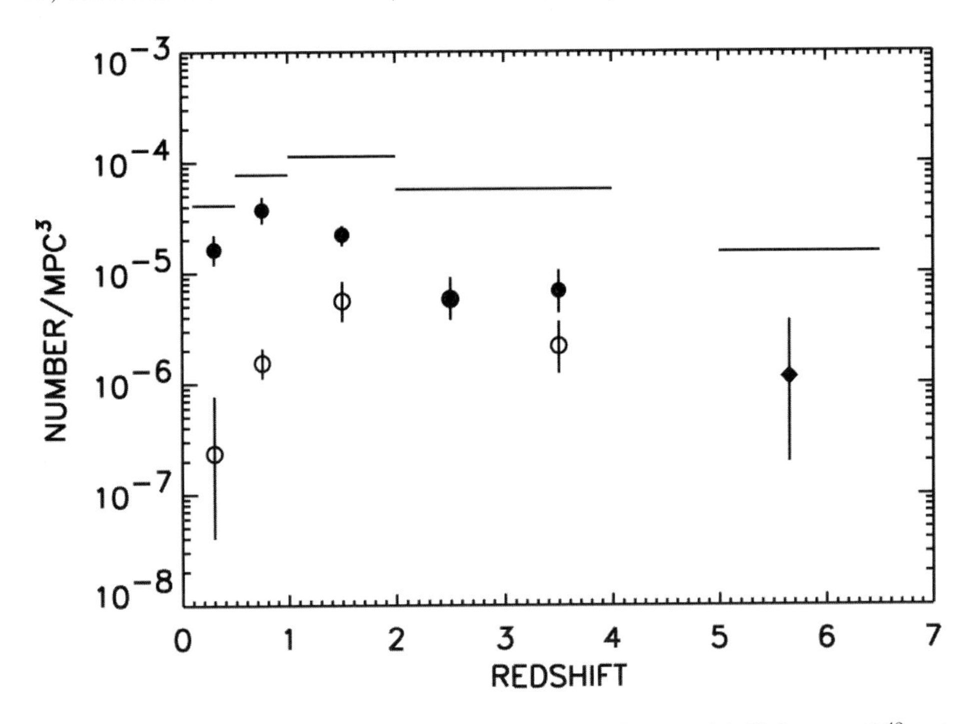

FIGURE 4: Comoving number density of likely AGN sources (L_X = 2–8 keV) between 10^{42} and $10^{44}\,\mathrm{erg\,s^{-1}}$ (filled circles), and between 10^{44} and $10^{45}\,\mathrm{erg\,s^{-1}}$ (open symbols). Upper horizontal lines provide maximal values from assigning all possible AGN to that redshift bin. Illustration from Barger et al. (2003).

Thus, the evidence for substantial luminosity evolution of QSOs, and number-evolution of narrow-line (Seyfert) galaxies, appears quite robust. Thus, we can say that the comoving density of QSOs beyond $z = 3$ declines strongly with z. The fundamental import of this finding is that re-ionization cannot have been dominated by AGN.

7.8 A Concluding Remark

It should be clear to readers of this book that our research efforts to directly observe and perhaps comprehend the basic phenomena and the varied nature of galaxy evolution in its early stages, has been only a partial success.

The follow-up science will likely contain surprises (it always seems to do so!); these adventures are the springboards to new understanding and the reason observers work long night hours at the telescope or in front of a computer terminal.

7.9 Acknowledgments

I wish to acknowledge the efforts of several colleagues and students for their contributions with specific results, ideas and comments on concepts.

In particular I thank Steve Dawson, Dan Stern, H. Yan, Rogier Windhorst, Mark Dickinson, Chuck Steidel, Martin White, and the "Subaru-team". We wish to thank Alice Shapley in particular, who read and gave comments on an earlier version of this book.

This book would never have been completed without the perseverance and attention of Dr Curtis Manning. Manning's effort was of particular import in Chapter 5. His overall role was that of an editorial aide and a co-investigator. I thank him for the substantial effort.

Finally, I am grateful to Clive Horwood of Praxis Publishing Ltd. for his perseverance against my easy distraction.

REFERENCES

Barger, A. J., et al., 2003, ApJL, 584, L61

Becker, R. H., et al., 2001, AJ, 122, 2850

Bell, E. F., et al., 2004, ApJ, 608, 752

Bouwens, R. J., et al., 2004, ApJL, 611, L1

Boyle, B. J., et al., 2000, MNRAS, 317, 1014

Bremer, M. N., et al., 2004, MNRAS, 347, L7

Cimatti, A., et al., 2004, Nature, 430, 184

Conselice, C. J., et al., 2003, AJ, 125, 66

Djorgovski, S. G., et al., 2001, ApJL, 560, L5

Drory, N., et al., 2004, ApJ, 608, 742

Ferguson, H. C., et al., 2004, ApJL, 600, L107

Fynbo, J. P. U., et al., 2003, AAp, 407, 147

Gnedin, N. Y. 2004, ApJ, 610, 9

Kogut, A., et al., 2003, ApJS, 148, 161

Lehnert, M. D. & Bremer, M. 2003, ApJ, 593, 630

Leitherer, C., et al., 1995, ApJL, 454, 19

Schmidt, M., et al., 1988, in ASP Conf. Ser. 2: *Optical Surveys for Quasars*, 87

Schmidt, M., et al., 1991, in ASP Conf. Ser. 21: *The Space Distribution of Quasars*, 109

Schneider, D. P., Set al., 1989, AJ, 98, 1507

Songaila, A. 2004, AJ, 127, 2598

Spergel, D. N., et al., 2003, ApJS, 148, 175

Spinrad, H., et al., 1998, AJ, 116, 2617

Steffen, A. T., et al., 2003, ApJL, 596, L23

Steidel, C. C., et al., 1999, ApJ, 519, 1

Steidel, C. C., et al., 2001, ApJ, 546, 665

White, R. L., et al., 2003, AJ, 126, 1

Yan, H. & Windhorst, R. A. 2004, ApJL, 612, L93

Index

Printing: Mercedes-Druck, Berlin
Binding: Stein + Lehmann, Berlin